Serono Symposia USA
Norwell, Massachusetts

Springer
New York
Berlin
Heidelberg
Barcelona
Budapest
Hong Kong
London
Milan
Paris
Santa Clara
Singapore
Tokyo

PROCEEDINGS IN THE SERONO SYMPOSIA USA SERIES

Continued after Index

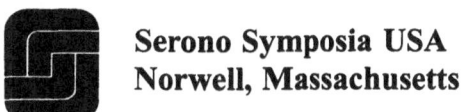

Serono Symposia USA
Norwell, Massachusetts

Barry R. Zirkin
Editor

Germ Cell Development, Division, Disruption and Death

With 56 Figures

Springer

Barry R. Zirkin, Ph.D.
Division of Reproductive Biology
Department of Population Dynamics
School of Hygiene and Public Health
Johns Hopkins University
Baltimore, MD 21205
USA

Proceedings of the XIVth Testis Workshop on Germ Cell Development, Division, Disruption and Death, sponsored by Serono Symposia USA, Inc., held February 19 to 22, 1997, in Baltimore, Maryland.

For information on previous volumes, please contact Serono Symposia USA, Inc.

Library of Congress Cataloging-in-Publication Data
Germ cell development, division, disruption and death / Barry R.
  Zirkin, editor.
      p. cm.
   "Proceedings of the XIVth Testis Workshop on Germ Cell Development, Division, Disruption and Death,
sponsored by Serono Symposia USA, Inc., held February 19–22, 1997, in Baltimore, Maryland"—T.p. verso.
   Includes indexes.

   ISBN-13:978-1-4612-7458-2   e-ISBN-13:978-1-4612-2206-4
   DOI:10.1007/978-1-4612-2206-4

   1. Spermatogenesis—Congresses. 2. Cell death—Congresses. I. Zirkin, B.R. II. Serono Symposia
USA. III. Testis Workshop on Germ Cell Development, Division, Disruption and Death (1997 : Baltimore, Md.)
   [DNLM: 1. Spermatozoa—cytology—congresses. 2. Spermatozoa—growth & development—con-
gresses. 3. Spermatogenesis—congresses. 4. Cell Death—congresses. WJ 834 G372 1998]
QL966.G475 1998
571.8'45—dc21   97-41289

Printed on acid-free paper.

© 1998 Springer-Verlag New York, Inc.
Softcover reprint of the hardcover 1st edition 1998

Production coordinated by Chernow Editorial Services, Inc., and managed by Francine McNeill; manufacturing
supervised by Joe Quatela.
Typeset by Agnew's, Inc., Grand Rapids, MI.

9 8 7 6 5 4 3 2 1

SPIN 10648965

XIVth TESTIS WORKSHOP ON GERM CELL DEVELOPMENT, DIVISION, DISRUPTION AND DEATH

Scientific Committee

Barry R. Zirkin, Ph.D., Chair
William J. Bremner, M.D., Ph.D.
Richard Bronson, M.D.
Claude Desjardins, Ph.D.
Maria L. Dufau, M.D., Ph.D.
E. Mitchell Eddy, Ph.D.
Mary Ann Handel, Ph.D.
Norman B. Hecht, Ph.D.
Marvin L. Meistrich, Ph.D.
Marie-Claire Orgebin-Crist, Ph.D.
David Phillips, Ph.D.
Bernard Robaire, Ph.D.
Patricia Saling, Ph.D.
Barbara M. Sanborn, Ph.D.
Jacquetta M. Trasler, M.D., Ph.D.

Organizing Secretary

Leslie Nies
Serono Symposia USA, Inc.
100 Longwater Circle
Norwell, Massachusetts

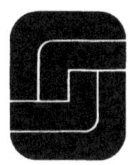

Preface

This book is composed of chapters prepared by invited speakers of the XIVth North American Testis Workshop, held in Baltimore, Maryland, February 19 to 22, 1997. The book reflects the desire of the Testis Workshop Scientific Committee for a program addressing a broad array of current research focused on the germ cell, from its development through its death. Consequently, the research of those who have contributed chapters to this book addresses, or can be applied to, the hormonal and molecular regulation of spermatogonial cell development and function (Nishimune, Orth, Russell, Dym), meiosis (Bickel, Wolgemuth, Handel, Trasler), spermatogenesis (Sassone-Corsi, Habener, Braun), cell death and aging (Campisi, Chen, Knudson, Swerdloff), and toxicant effects on spermatogenesis (Sharpe, Kelce, Robaire, Meistrich).

In their preface to the proceedings of the 1988 Testis Workshop, Drs. Larry Ewing and Bernard Robaire noted that the first published description of seminiferous tubules in the testis was in 1668, by C. Aubry, and that "the exciting papers in these proceedings [of the 1988 Workshop] emphasize the significant progress that has been made since the mid-17th century in understanding testis and function." It is quite clear that current research is proceeding at a breathtaking pace, with enormous strides having been made even since the XIIIth Workshop in 1995. If the XIVth Workshop succeeded, and this book succeeds, in conveying this excitement in our field (which, of course, was our intent), credit belongs to the creativity of the Scientific Committee, whose members were William J. Bremner, Richard Bronson, Claude Desjardins, Maria L. Dufau, E. Mitchell Eddy, Mary Ann Handel, Norman B. Hecht, Marvin L. Meistrich, Marie-Claire Orgebin-Crist, David Phillips, Bernard Robaire, Patricia Saling, Barbara M. Sanborn, and Jacquetta M. Trasler.

The North American Testis Workshop began in 1972, with its primary support then coming from NIH. Remarkably, through the ensuing 25 years, the Workshop has consistently represented the cutting edge of male reproduction. The excellence of the Workshop, and, indeed, its very existence, depend heavily on the financial support that it attracts. In that regard, the role of Serono Symposia USA in the Workshop cannot be overemphasized. Since 1993, Serono Symposia USA has become increasingly involved in the Workshop; and by 1997, Serono Symposia USA, under the leadership of Leslie Nies, President, took on its full support. That support has been far more than financial; for the 1997 meeting, all conference arrangements were handled flawlessly by Dianne Ferreira, and all publications, including this book, were produced under the outstanding leadership of Judy Donahue. On behalf of all of us who care about excellent science, I wish to express my gratitude to Leslie Nies and to her staff for their confidence in us and for their support. They have allowed the scientists the freedom to do what scientists do best—science—while seeing to all the important details that make or break a meeting. Personally, I cannot thank Dianne Ferreira and Judy Donahue enough for making my job so very easy, and so enjoyable. With the involvement of Serono Symposia USA, I am confident that the Workshop will remain at the cutting edge of our field for years to come.

BARRY R. ZIRKIN

Contents

**Part I. Spermatogonial Cell Development, Function,
and Regulation**

Part V.   Toxicant Effects on Spermatogenesis

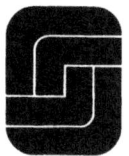

Contributors

VALERIE BESSET, Department of Genetics and Development, Columbia University College of Physicians and Surgeons, New York, New York, USA.

SHARON E. BICKEL, Whitehead Institute for Biomedical Research, Cambridge, Massachusetts, USA.

ROBERT E. BRAUN, Department of Genetics, University of Washington, Seattle, Washington, USA.

RALPH L. BRINSTER, School of Veterinary Medicine, University of Pennsylvania, Philadelphia, Pennsylvania, USA.

JUDITH CAMPISI, Department of Cancer Biology, Berkeley National Laboratory, Berkeley, California, USA.

HAOLIN CHEN, Department of Population Dynamics, Johns Hopkins University School of Hygiene and Public Health, Baltimore, Maryland, USA.

JOHN COBB, Department of Biochemistry and Cellular and Molecular Biology, University of Tennessee, Knoxville, Tennessee, USA.

PHILIP B. DANIEL, Laboratory of Molecular Endocrinology, Howard Hughes Medical Institute, Massachusetts General Hospital, Boston, Massachusetts, USA.

TONIA E. DOERKSEN, Departments of Pediatrics, Human Genetics, and Pharmacology and Therapeutics, McGill University, and The McGill University-Montreal Children's Hospital Research Institute, Montreal, Quebec, Canada.

MARTIN DYM, Department of Cell Biology, Georgetown University Medical Center, Washington, D.C., USA.

JANE S. FISHER, Medical Research Council (MRC) Reproductive Biology Unit, Centre for Reproductive Biology, Edinburgh, Scotland, UK.

JOEL F. HABENER, Laboratory of Molecular Endocrinology, Howard Hughes Medical Institute, Massachusetts General Hospital, Boston, Massachusetts, USA.

BARBARA F. HALES, Department of Pharmacology and Therapeutics, McGill University, Montreal, Quebec, Canada.

MARY ANN HANDEL, Department of Biochemistry and Cellular and Molecular Biology, University of Tennessee, Knoxville, Tennessee, USA.

AMIYA SINHA HIKIM, Endocrinology/Medicine, Harbor-UCLA Medical Center, Torrance, California, USA.

WILLIAM P. JESTER, Department of Anatomy and Cell Biology, Temple University School of Medicine, Philadelphia, Pennsylvania, USA.

MARKO KANGASNIEMI, Department of Experimental Radiation Oncology, University of Texas M.D. Anderson Cancer Center, Houston, Texas, USA.

WILLIAM R. KELCE, Reproductive Toxicology Division, Endocrinology Branch, National Health and Environmental Effects Research Laboratory, U.S. Environmental Protection Agency, Research Triangle Park, and The Laboratories for Reproductive Biology and the Department of Pediatrics, University of North Carolina, Chapel Hill, North Carolina, USA.

JEFFREY B. KERR, Department of Anatomy, Monash University, Clayton, Victoria, Australia.

C. MICHAEL KNUDSON, Departments of Medicine and Pathology, Washington University School of Medicine, St. Louis, Missouri, USA.

STANLEY J. KORSMEYER, Departments of Medicine and Pathology, Washington University School of Medicine, St. Louis, Missouri, USA.

LING-HONG LI, Department of Anatomy and Cell Biology, Temple University School of Medicine, Philadelphia, Pennsylvania, USA.

DONG LIU, Department of Genetics and Development, Columbia University College of Physicians and Surgeons, New York, New York, USA.

YANHE LUE, Endocrinology/Medicine, Harbor-UCLA Medical Center, Torrance, California, USA.

LINDI LUO, Department of Population Dynamics, Johns Hopkins University School of Hygiene and Public Health, Baltimore, Maryland, USA.

GREGOR MAJDIC, Medical Research Council (MRC) Reproductive Biology Unit, Centre for Reproductive Biology, Edinburgh, Scotland, UK.

MARVIN L. MEISTRICH, Department of Experimental Radiation Oncology, University of Texas M.D. Anderson Cancer Center, Houston, Texas, USA.

CARMEN MERTINEIT, Departments of Pediatrics, Human Genetics, and Pharmacology and Therapeutics, McGill University, and The McGill University-Montreal Children's Hospital Research Institute, Montreal, Quebec, Canada.

MIKE R. MILLAR, Medical Research Council (MRC) Reproductive Biology Unit, Centre for Reproductive Biology, Edinburgh, Scotland, UK.

LUCIA MONACO, Institute of Genetics and of Molecular and Cellular Biology, National Center of Scientific Research, Illkirch, Strasbourg, France.

YOSHITAKE NISHIMUNE, Science for Laboratory Animal Experimentation, Research Institute for Microbial Diseases, Suita, Osaka, Japan.

TERRY L. ORR-WEAVER, Whitehead Institute for Biomedical Research, Cambridge, Massachusetts, USA.

JOANNE M. ORTH, Department of Anatomy and Cell Biology, Temple University School of Medicine, Philadelphia, Pennsylvania, USA.

PRIYANKE PARTE, Medical Research Council (MRC) Reproductive Biology Unit, Centre for Reproductive Biology, Edinburgh, Scotland, UK.

JIANPING QIU, Department of Experimental Pathology, Holland Laboratory, Rockville, Maryland, USA.

TRIPATHI RAJAVASHISTH, Endocrinology/Medicine, Harbor-UCLA Medical Center, Torrance, California, USA.

NEELAKANTA RAVINDRANATH, Department of Cell Biology, Georgetown University Medical Center, Washington, D.C., USA.

KUNSOO RHEE, Department of Genetics and Development, Columbia University College of Physicians and Surgeons, New York, New York, USA.

BERNARD ROBAIRE, Department of Pharmacology and Therapeutics, McGill University, Montreal, Quebec, Canada.

LONNIE D. RUSSELL, Department of Physiology, School of Medicine, Southern Illinois University, Carbondale, Illinois, USA.

PAOLO SASSONE-CORSI, Institute of Genetics and of Molecular and Cellular Biology, National Center of Scientific Research, Illkirch, Strasbourg, France.

PHILIPPA T.K. SAUNDERS, Medical Research Council (MRC) Reproductive Biology Unit, Centre for Reproductive Biology, Edinburgh, Scotland, UK.

RICHARD M. SHARPE, Medical Research Council (MRC) Reproductive Biology Unit, Centre for Reproductive Biology, Edinburgh, Scotland, UK.

RONALD S. SWERDLOFF, Endocrinology/Medicine, Harbor-UCLA Medical Center, Torrance, California, USA.

HIROMITSU TANAKA, Science for Laboratory Animal Experimentation, Research Institute for Microbial Diseases, Suita, Osaka, Japan.

JACQUETTA M. TRASLER, Departments of Pediatrics, Human Genetics, and Pharmacology and Therapeutics, McGill University, and The McGill University-Montreal Children's Hospital Research Institute, Montreal, Quebec, Canada.

KENNETH S.K. TUNG, Department of Pathology, University of Virginia Health Sciences Center, Charlottesville, Virginia, USA.

KATIE J. TURNER, Medical Research Council (MRC) Reproductive Biology Unit, Centre for Reproductive Biology, Edinburgh, Scotland, UK.

WILLIAM H. WALKER, Department of Cell Biology and Physiology, University of Pittsburgh School of Medicine, Pittsburgh, Pennsylvania, USA.

CHRISTINA WANG, Endocrinology/Medicine, Harbor-UCLA Medical Center, Torrance, California, USA.

ELIZABETH M. WILSON, The Laboratories for Reproductive Biology, Departments of Pediatrics and Biochemistry and Biophysics, University of North Carolina, Chapel Hill, North Carolina, USA.

GENE WILSON, Department of Experimental Radiation Oncology, University of Texas M.D. Anderson Cancer Center, Houston, Texas, USA.

DEBRA J. WOLGEMUTH, Department of Genetics and Development, Columbia University College of Physicians and Surgeons, New York, New York, USA.

QI ZHANG, Department of Genetics and Development, Columbia University College of Physicians and Surgeons, New York, New York, USA.

BARRY R. ZIRKIN, Department of Population Dynamics, Johns Hopkins University School of Hygiene and Public Health, Baltimore, Maryland, USA.

Part I

Spermatogonial Cell Development, Function, and Regulation

1

Cloning and Characterization of Genes Specifically Expressed in Germ Line Cells

HIROMITSU TANAKA AND YOSHITAKE NISHIMUNE

Spermatogenesis is a complicated process involving many interesting phenomena that are biologically important, such as stem cell proliferation and differentiation, meiosis, generation of haploid germ cells, and morphogenesis of the developing sperm. To understand these processes, we have over a long period of time studied the testis from various aspects using many different analytical methods. However, there appears to be still a long way to go for us to obtain clear understanding of this mysterious organ.

The initial history of testicular biology drawn from morphological observations indicates that many drastic morphological changes occur during spermatogenesis. With the development of many varied biochemical studies, we began to realize that large numbers of specific molecules are involved in the morphological changes occurring in spermatogenesis, and that it is difficult to clarify the relationships between the specific molecules and their morphological and functional roles. However, with the help of immunological and biochemical techniques, identification and isolation of the specific molecules became easier. Many novel molecules have been isolated and characterized in this decade by using immunological techniques for monoclonal and specific polyclonal antibodies.

Furthermore, recent progress in molecular biology has stimulated the investigation for deeper understanding of many complicated biological processes. Isolation of genes specifically expressed in testes and also specifically expressed in different steps of germ cell development has shed new light on understanding spermatogenesis. Progress in molecular biological techniques has been continually instrumental for understanding the regulation of the specific molecules in germ cell differentiation, the relationships between gene products and their physiological roles, and the morphological changes in cell differentiation.

Molecular embryology and embryological technology also helped us analyze the function of genes involved in cell differentiation and normal development of

3

embryos. Introducing specific genes into early embryos (transgenic animals), we are able to analyze the function of genes as well as the regulatory mechanisms of specific gene expression. The technique of gene targeting to produce knockout animals also became a powerful methodology for the study of in vivo gene functions.

Combining these techniques, new knowledge regarding spermatogenesis is now being acquired, accumulating day by day, as is seen in other fields of biology.

Identification of Specific Molecules in the Steps of Spermatogenesis

Many specific molecules involving germ cell differentiation have been identified by standard biochemical methods. The use of some lectins are more effective for isolation of specific receptor molecules on the surface of germ cells (1, 2). However, it is much more powerful to use specific antibodies to identify the germ cell-specific molecules. As is seen in the case of autoallergic orchitis or frequent induction of sperm-specific antibodies in both males and females, testicular germ cells are unique and different from other somatic cells from the immunological point of view. As the germ cells are separated from the immunological surveillance mechanism (3), the antibodies capable of reacting with germ cells are easily induced when testicular germ cells are immunized. Furthermore, the specific antibodies are induced to react exclusively with testicular germ cells when immunized to isogenic or autogenic animals. These antibodies could recognize many kinds of antigen molecules that appear or disappear in association with germ cell differentiation. Furthermore, monoclonal antibodies recognizing germ cells could also be isolated. These monoclonal antibodies are extending our study into the elucidation of specific expression and characterization of the antigen molecules (4–8). Table 1.1 shows some of the monoclonal antibodies isolated in our laboratory.

TABLE 1.1. Testis-specific antigens detected by monoclonal antibodies.

Monoclonal antibody	Antigen	(MW, pl)	Expression	References
BC 7	TDA95	(95 KDa)	zyg–pach	Koshimuzu et al. 1993 (1)
CA 12	TDA 95	(95 KDa)		
EE 2	TDA 114	(114 KDa, 6.1)	A gonia–zyg	Koshimuzu et al. 1993
TRA 369	Calmegin	(95 KDa, 5.2)	pach–round spermatid	Watanabe et al. 1992, 1994
TRA 104	Gena 110	(110 KDa, 7.1)	PGC–spermatid	Unpublished observations
TRA 7	SLA 200	(<200 KDa)	spermatid	Unpublished observations

Germ cell-specific monoclonal antibodies isolated in our laboratory. These antigens were expressed stage specifically in germ cells during spermatogenesis. Zyg, zygotene spermatocyte; pach, pachytene spermatocyte; A gonia, type A spermatogonia; PGC, primordial germ cells.

Molecular Cloning of Genes Specifically Expressed in Testicular Germ Cells

It would be useful to isolate germ cell-specific cDNA clones and genomic DNAs for further study of some molecules specifically expressed in germ cell differentiation. However, no systematic isolation of germ cell-specific cDNA clones has been carried out, although several genes known to be germ cell-specific genes have already been isolated (9). Recently, some new systematic approaches for the study of germ cell differentiation have successfully isolated novel genes that are germ cell specific (10).

Isolation of Genes Coding for the Antigen Molecules Specifically Expressed in Testicular Germ Cells

Many monoclonal antibodies recognizing specific germ cells could be useful for isolation of the cDNAs. Some of them, however, did not necessarily work well. In many cases, epitopes for monoclonal antibodies are not on polypeptides but on the sugar moieties of glycoproteins. By using these monoclonal antibodies we are not able to isolate cDNA clones from expression libraries of bacteriophage in *E. coli,* because no modification of encoded proteins by sugar moieties can occur in bacteria. In other cases, the avidity of those antibodies is not strong enough to isolate their corresponding cDNA clones from expression libraries. In these cases, isolation of cDNAs is not easy but needs one more step to obtain another antibody capable of recognizing an epitope of the polypeptide molecule or polyclonal antibodies with strong affinity. In any case, this approach has been successful for isolation of novel genes specifically expressed in testicular germ cells (11).

Testicular germ cells are easily recognized by immunological surveillance, and upon immunization with testicular germ cells the immunized animals respond to produce many types of antibodies. If we immunize rabbits with mouse testicular germ cells, we can induce many types of antibodies specifically reacting with germ cells as well as antibodies generally reacting with mouse cells. To obtain specific antibodies that only react with germ cell-specific molecules, the rabbit antiserum is absorbed intraperitoneally by injecting the antiserum into a mouse whose testes have been previously castrated (in vivo absorption). Bleeding the mouse, we can obtain specific rabbit antibodies that react with all the types of mouse germ cells but not with other mouse somatic cells (12). The germ cell-specific molecules that reacted with this rabbit antiserum are shown in Figure 1.1. Using such monoclonal and polyclonal antibodies, we can isolate cDNA clones coding for specific antigenic molecules (11).

Subtraction Cloning for Isolation of Testicular Germ Cell–Specific cDNA Clones

When antibody is not effectively produced for some reason, such as a small number of antigen molecules or low antigenicity, we cannot isolate the corre-

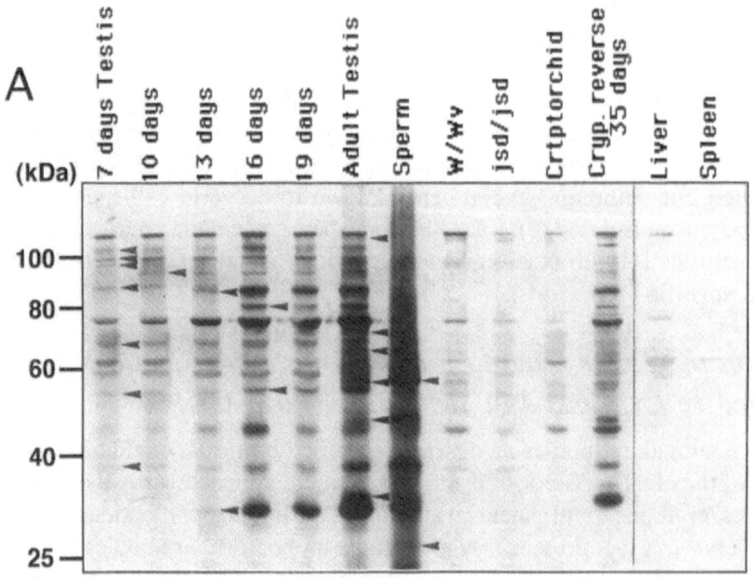

FIGURE 1.1. Immunoblot analysis of antigenic molecules in the testis (A) and schematic representation of testicular germ cell-specific antigens during spermatogenesis (B). Proteins extracted from testes of 7-day-old, 10-day-old, 13-day-old, 16-day-old, 19-day-old, adult, W/W^v and jsd/jsd mice, from cryptorchid testes, and testes 35 days after the surgical reversal of the cryptorchid condition, and from the liver, spleen, and sperm, were separated by sodium dodecyl sulfate-polyacrylamide gel electrophoresis (SDS-PAGE), transferred to a filter, and immunostained with the serum. Arrowheads indicate antigens appearing first during testicular germ cell differentiation. Molecular weights are noted at the left margin (A). To show expression patterns of the germ cell-specific antigens, some parts of the results are schematically depicted (B). Detection of each antigen is indicated by a solid bar.

sponding cDNA clones (i.e., regulatory proteins are present in smaller amounts than structural proteins). Thus, it is difficult to prepare sufficient amounts to induce antibody production. In this situation, to isolate these cDNA clones, subtraction cloning would be effective (13, 14). No germ cell differentiation occurs in the testes of some types of mutant mice (W/W^v) even though the supporting cells of the testes apparently maintain normal function. If we can subtract a cDNA library of such a mutant testis from that of wild-type animals, cDNA clones derived solely from the germ cells at all steps of differentiation can be isolated. Because of technical improvement for preparing this type of subtraction cDNA library, many cDNA clones specifically expressed in testicular germ cells have been isolated (14). Figure 1.2 shows the screening of the testis-specific genes in

Probe: W/W^v mutant testis

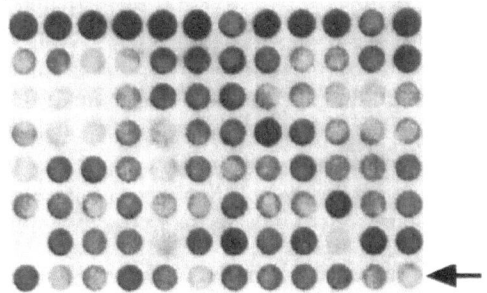

Probe: wild type testis

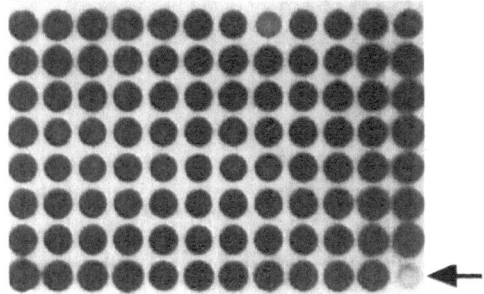

FIGURE 1.2. Dot blot analysis of cDNA clones randomly selected from the subtracted cDNA library. Plasmid DNA (10 ng) from each clone obtained from the subtracted cDNA library was applied onto a nylon membrane using the convertible filtration manifold system apparatus (BRL) and hybridized with digoxigenin- (DIG-) labeled RNA synthesized in vitro from the mutant (A) and wild-type (B) testicular library, respectively. The overlined 6 dots represent control cDNA clones randomly picked from the nonsubtracted wild-type testicular library, and the solid arrows mark dots of pAP3neo vector without an insert as a negative control.

the subtracted cDNA library. Use of various mutant mice and application of such techniques likely lead us to isolate specific cDNA clones expressed only in spermatogonia, spermatocytes, or spermatids.

Analysis of the Function of Proteins Encoded by cDNAs and the In Vitro Expression of the Clones

In vitro culture systems are useful for investigation of the regulation of isolated genes and the function of encoded products. However, we do not yet have any efficient culture cell lines or in vitro cultivation systems for investigating the function of germ cell-specific genes. In some cases, culture cell lines could be useful, such as ES (embryonic stem), EG (embryonic germ), and EC (embryonal carcinoma) cell lines (15–17), because some of the germ cell-specific genes are expressed in these cell lines. Co-cultivation of testicular germ cells with some Sertoli cell lines can induce differentiation of the germ cells, proceeding to the meiotic prophase and then to production of haploid germ cells (18). Use of such a transforming Sertoli cell line may be effective to analyze the regulation of germ cell-specific genes. However, this system does not induce the complete germ cell differentiation and is not sufficient for further analysis. Thus, we must develop a new cultivation system or culture cell line of testicular germ cells to analyze the expression of germ cell-specific genes.

Analysis of the Regulation of Gene Expression in Transgenic Animals

As discussed previously, in vitro cultivation of the germ cells is technically premature to study the regulation of gene expression. To overcome the problem of premature in vitro cultivation system, transgenic animals can be used. To obtain transgenic animals, exogenous DNA is introduced into fertilizing eggs. The exogenous DNA is made with the DNA having a 5'-noncoding sequence of a certain specific gene fused with a reporter gene such as the chloramphenicol acetyl transferase (CAT) gene. If the specific expression of CAT is observed with an appropriate sequence of the 5'-noncoding region, then we are able to obtain the information on the *cis* element of germ cell-specific and also differentiation-specific expression (19). Analyses of transgenic mice indicate some of the testis-specific genes do not have TATA and CAAT boxes but GC-rich regions. In some cases, the regulation of exogenous gene expression does not always reflect the regulation of physiological gene expression because the copy number of a transgene and its integration sites affect the expression of the gene. Therefore, we have to raise many lines of transgenic mice before making a conclusion from the results (20).

Gene Targeting Study on the Function of Specific Genes Isolated from Testicular Germ Cells

In many cases it is difficult to clarify by biochemical analysis the function of product proteins encoded by the cDNA. Even when we successfully identify biochemical functions of certain specific proteins in vitro, this does not mean that we understand the real role of the gene products in vivo. Analyses of defective genes resulting from some mutations lead us to understand the gene functions, and thus it would be useful to generate artificial mutant animals. By the development of ES cell lines that exhibit excellent germ cell transmission, gene targeting became a powerful technique.

The calmegin gene was cloned and analyzed in our laboratory. Calmegin protein, which is expressed specifically in meiosis, functions as a molecular chaperone. The physiological function of this molecule in germ cell differentiation was studied by targeted disruption of its gene. The calmegin knockout male mouse showed infertility in loss of sperm adhesion to the egg, indicating that calmegin functions as a chaperone for sperm-surface proteins to mediate sperm–egg interactions (unpublished observations).

Many genes are targeted, and we have many knockout mouse strains. However, some of them are embryolethal in the homozygous situation and it is therefore difficult to identify the function of knocked-out genes. Even in this kind of situation, conditional knockout of the gene could be effectively accomplished. The development of the cre-loxP recombination system facilitates the study of the effect of gene targeting exclusively in some specific conditions (21). By providing options of where and when to knock out the gene, this technique allows us to isolate the knockout mouse even when it is embryo lethal.

By further development and progression of the techniques discussed here, we will surely make great progress in the study of spermatogenesis in the near future. In turn, detailed knowledge of germ cell proliferation and differentiation will facilitate developing new techniques, even new clinical applications, that will control the complicated processes of germ cell differentiation.

Acknowledgment. We thank Dr. A. Tanaka for reviewing this manuscript.

References

1. Millette CF, Scott BK. Identification of spermatogenic cell plasma membrane glycoproteins by two-dimensional electrophoresis and lectin blotting. J Cell Sci 1984;65:233–48.
2. Schopperle WM, Armant DR, Dewolf WC. Purification of a tumor-specific PNA-binding glycoprotein, gp200, from a human embryonal carcinoma cell line. Arch Biochem Biophys 1992;298:538–43.
3. Welber JE, Turner TT, Tung KSK, Russell LD. Efferct of cytochalasin D on the integrity of the Sertoli cell (blood-testis) barrier. Am J Anat 1988:182:130–47.

4. Hahnel AC, Eddy EM. Cell surface marker of mouse primordial germ cells defined by two monoclonal antibodies. Gamete Res 1986;15:25–34.
5. Fenderson BA, O'Brien DA, Millette CF, Eddy EM. Stage-specific expression of three cell surface carbohydrate antigens during murine spermatogenesis detected with monoclonal antibodies. Dev Biol 1984;103:117–28.
6. Koshimizu U, Watanabe D, Sawada K, Nishimune Y. A novel stage-specific differentiation antigen is expressed on mouse testicular germ cells during early meiotic prophase. Biol Reprod 1993;49:875–84.
7. Koshimizu U, Nishioka H, Watanabe D, Dohmae K, Nishimune Y. Characterization of a novel spematogenic cell antigen specific for early stage of germ cells in mouse testis. Mol Reprod Dev 1995;40:221–7.
8. Watanabe D, Sawada K, Koshimizu U, Kagawa T, Nishimune Y. Characterization of male meiotic germ cell-specific antigen (Meg 1) by monoclonal antibody TRA369 in mice. Mol Reprod Dev 1992;33:307–12.
9. Wolgemuth DJ, Watrin F. List of cloned mouse genes with unique expression patterns during spermatogenesis. Mamm Genome 1991;1:283–8.
10. Yuan L, Liu JG, Hoog C. Rapid cDNA sequencing in combination with RNA expression studies in mice identifies a large number of male germ cell-specific sequence tags. Biol Reprod 1995;52:131–8.
11. Watanabe D, Yamada K, Nishina Y, Tajima Y, Koshimizu U, Nagata A, Nishimune Y. Molecular cloning of a novel $Ca(2^+)$-binding protein (calmegin) specifically expressed during male meiotic germ cell development. J Biol Chem 1994;269:7744–9.
12. Tsuchida J, Nishina Y, Akamatsu T, Nishimune Y. Characterization of development-specific, cell type-specific mouse testicular antigens using testis-specific polyclonal antibodies. Int J Androl 1995;18:208–12.
13. Starborg M, Brundell E, Hoog C. Analysis of the expression of a large number of novel genes isolated from mouse prepubertal testis. Mol Reprod Dev 1992;33:243–51.
14. Tanaka H, Yoshimura Y, Nishina Y, Nozaki M, Nojima H, Nishimune Y. Isolation and characterization of cDNA clones specifically expressed in testicular germ cells. FEBS Lett 1994;355:4–10.
15. Doetschman TC, Eistetter H, Katz M, Schmidt W, Kemler R. The in vitro development of blastocyst-derived embryonic stem cell lines: formation of visceral yolk sac, blood islands and myocardium. J Embryol Exp Morphol 1985;87:27–45.
16. Matsui Y, Zsebo K, Hogan BL. Derivation of pluripotential embryonic stem cells from murine primordial germ cells in culture. Cell 1992;70:841–7.
17. Martin GR, Evans MJ. The morphology and growth of a pluripotent teratocarcinoma cell line and its derivatives in tissue culture. Cell 1974;2:163–72.
18. Rassoulzadegan M, Paquis-Flucklinger V, Bertino B, Sage J, Jasin M, Miyagawa K et al. Transmeiotic differentiation of male germ cells in culture. Cell 1993;75:997–1006.
19. Watanabe D, Okabe M, Hamajima N, Morita T, Nishina Y, Nishimune Y. Characterization of the testis-specific gene 'calmegin' promoter sequence and its activity defined by transgenic mouse experiments. FEBS Lett 1995;365:509–12.
20. Zambrowicz BP, Harendza CJ, Zimmermann JW, Brinster RL, Palmiter RD. Analysis of the mouse protamine 1 promoter in transgenic mice. Proc Natl Acad Sci USA 1993;90:5071–5.
21. Barinaga M. Knockout mice: round two. Science 1994;265:26–8

2

Gonocyte–Sertoli Cell Interactions in Testes of Neonatal Rats

JOANNE M. ORTH, WILLIAM P. JESTER, JIANPING QIU, AND LING-HONG LI

Our understanding of spermatogenesis in the adult testis has increased substantially in past years, as more and more investigators have focused on the role of interactions between somatic and germ cells in supporting this process. We now recognize, for example, the way in which adhesive interactions between Sertoli and germ cells control movement of young spermatocytes from the basal into the adluminal compartment and the critical support provided to the adluminal cohort of spermatogenic cells by the Sertoli cells. Considerably less is known, however, about how spermatogonia begin their passage through spermatogenesis and even less is understood about the ancestors of these cells, the gonocytes, and how they mature during the prespermatogenic period of testicular development.

In our laboratory, we have been investigating development of neonatal Sertoli and germ cells of rats [see (1) for review], with the ultimate aim of identifying the events that ensure successful onset of spermatogenesis and understanding how these events are regulated. We have found that, while the proliferative activity of Sertoli cells is declining with increasing postnatal age, gonocytes reinitiate mitosis on the third day after birth and, shortly thereafter, migrate actively to reach the periphery of the seminiferous cord (2–4). It is at this new location that they encounter extracellular matrix factors in the basal lamina and are thought to mature into the first generation of type A spermatogonia. We have established a Sertoli–gonocyte co-culture system (5) that mimics closely the relationship between these two cell types in vivo and in which Sertoli cells adhere via laminin (6) and possibly other matrix molecules to the substrate and in which gonocytes apparently mature normally (5). We have largely utilized this system to investigate the ways in which Sertoli cells and gonocytes of neonates interact during this critical developmental period. In this chapter, we summarize some of our recent findings in this regard.

Neural Cell Adhesion Molecule (NCAM) and
Sertoli Cell–Gonocyte Adhesion

In the testis of the newborn rat, gonocytes are located at some distance from the basal lamina and are thus totally enclosed by Sertoli cells. By postnatal day 4, however, some gonocytes develop peripherally directed pseudopodial processes and begin to migrate to the periphery (4). In this way, these cells move into position in anticipation of later formation of the basal compartment. This change in position is apparently vital for gonocyte survival because any cells that remain centrally located eventually die (7). Observations made both in vivo (3) and in vitro (5) indicate that, during their migratory period, gonocytes adhere avidly to Sertoli cells, which constitute the substrate upon which they move. Thus, adhesive mechanisms that operate between these two cell types are doubtless of vital importance to gonocyte migration.

We have investigated Sertoli cell–gonocyte adhesion in vitro with several approaches, and, in initial studies, found that gentle trypsinization disassociates gonocytes from Sertoli cells regardless of the presence or absence of Ca^{2+} in the medium; however, we also found that adhesion between these cells is unaffected by reduction of the incubation temperature to 4°C (8). These findings suggest that cadherins are not important in Sertoli cell–gonocyte adhesion. Subsequently, we used immunolocalization to identify NCAM at Sertoli–Sertoli and Sertoli–gonocyte interfaces (Fig. 2.1). Finally, we found that incubation of Sertoli–gonocyte co-cultures with anti-NCAM shortly after plating caused detachment of germ cells from the underlying Sertoli monolayer, thus verifying this factor's functional role in adhesion of gonocytes to Sertoli cells (8). Interestingly, although NCAM is present at Sertoli–Sertoli interfaces, these cells retained their attachment to the underlying Matrigel substrate throughout our manipulations, suggesting that another mechanism, possibly integrin based, is critical for interaction between these cells and extracellular matrix and is sufficient to maintain the monolayer.

NCAM is a pleiomorphic molecule whose potential for serving as a regulatory factor in cell–cell interaction is well recognized [for review, see (9)]. For example, the presence or absence of extracellular polysialic acid (PSA) is believed to affect not only the degree of NCAM-based adhesiveness between cells but also to regulate general spacing between cells and, as a result, to affect the way in which some ligands can interact with their surface receptors. Thus, NCAM has the potential to regulate other ligand-based modifiers of gonocyte development. In our studies, we have found that NCAM in neonatal testes is uniformly PSA negative (8), which is consistent with the evidence for tight adhesiveness between Sertoli cells and postnatal gonocytes. This finding supports the notion that, besides serving adhesiveness between these cells, NCAM facilitates other contact-mediated forms of interaction between them.

In addition, NCAM is a pleiomorphic molecule found in several isoforms, typically 120, 140, or 180 kDa. Moreover, the form(s) expressed by a given cell may reflect the way in which the factor affects cell function. For example, the

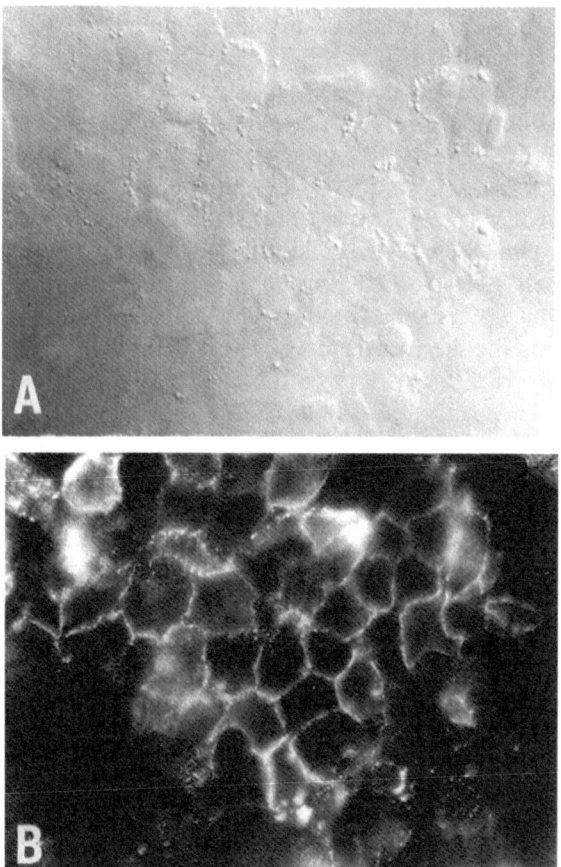

FIGURE 2.1. Immunofluorescent localization of neural cell adhesion molecule (NCAM) at cellular surfaces in a co-culture of neonatal Sertoli cells and gonocytes. (*A*) Differential interference contrast. (*B*) Epifluorescence. This culture was exposed to primary antiserum recognizing all isoforms of NCAM, followed by rhodamine-conjugated secondary antibodies. Controls lacking primary antiserum were uniformly negative. Adapted with permission from Orth and Jester (8).

lighter, 120-kDa form is a peripheral membrane protein that lacks the ability to interact with cytoskeletal elements, while the larger forms may participate in actin-based intercellular linkages (9) and thus have the potential to affect expression of various genes by a cell. For this reason, we have begun to study the nature and function of NCAM in developing testes in vivo, with the aim of determining whether it has a role in determining the fate of individual gonocytes or Sertoli cells. Our initial findings from a Western analysis with protein isolated from testes of rats from the day of birth through adulthood indicate that neonatal testes express the 140-kDa form exclusively from birth through day 15, and that, between days 5 and 15, there is a dramatic drop in the amount of NCAM synthesized (Orth, unpublished data). This may reflect the changing cellular makeup of the seminiferous epithelium as spermatogenesis begins. After day 15, NCAM essentially disappears from the testis (Orth, unpublished data), to be replaced by another, possibly cadherin-based (10) adhesive mechanism. From our analysis thus far, we know that (1) testicular NCAM is uniformly PSA negative and (2) the 140-kDa isoform of NCAM is expressed by testes at increasingly lower levels through postnatal day 15. Our findings to date identify NCAM as important for maintenance of Sertoli cell—Sertoli cell and Sertoli cell—gonocyte adhesion during neonatal through prepuberal testicular development in the rat. Ongoing studies will be aimed at identifying the specific role of NCAM in cell—cell interactions in maturing rat testis.

The Role of *c-kit* in Gonocyte Maturation

The *c-kit* proto-oncogene encodes for a plasma membrane-associated receptor with inherent tyrosine kinase activity. This receptor and its ligand, stem cell factor (SCF), are vital for normal hematopoiesis, melanogenesis, and gametogenesis. Substantial evidence also indicates that SCF is manufactured by Sertoli cells and that *c-kit* is expressed by germ cells during fetal development and in puberal and adult rats [for review, see (11)]. It has been suggested that differentiating spermatogonia may be SCF dependent while undifferentiated spermatogonia are not (12, 13), and that the role of the *c-kit*/SCF system among responsive germ cell populations, at least in pubertal/adult animals, is to control loss of these cells via apoptosis (14). Because of the many indications that interactions between neonatal Sertoli cells and gonocytes are of critical importance to development of the latter and to eventual onset of spermatogenesis, we have begun to ask whether expression of *c-kit* by gonocytes and SCF by Sertoli cells in neonates has any role in development of one or both of these cell types. Our observations thus far are summarized next.

In recently published initial studies (15), we asked whether *c-kit* is expressed in neonatal gonocytes and, if so, what might be its functional role. We isolated total RNA from decapsulated testes collected from 1- to 5-day-old pups and from adults and probed for expression of *c-kit* with Northern analysis, using a ^{32}P-labeled probe produced by random priming, with a fragment spanning nucleotides 2607—3710 of the murine *c-kit* gene as template. With this analysis, we detected a major transcript of approximately 5.5 kb in all three samples, confirming expression of this gene in both neonates and adults. Subsequent Western analysis, using an affinity-purified polyclonal rabbit anti-*c-kit* antibody, confirmed synthesis of the *c-kit* receptor protein in these testes. To identify precisely the cells expressing this gene in neonates, and to confirm suggestions by others that spermatogonia of adults are *c-kit* positive (16), we also carried out in situ hybridizations and immunolocalizations to identify transcripts and receptor protein, respectively, in cells of neonatal and adult testes. For in situ hybridizations, we utilized digoxigenin-labeled cRNA sense and antisense probes prepared by in vitro transcription of the same probe used for Northern analyses. Our findings (Fig. 2.2) indicate that (i) essentially all gonocytes express this gene, at varying levels of intensity, on the day of birth just before onset of their migration and reinitiation of mitosis (3); (ii) by postnatal day 5, when these cells are actively migrating and some are proliferating (3, 4), strong signal indicating the presence of *c-kit* transcripts is seen in essentially all gonocytes; and (iii) in testes of adults, spermatogonia and possibly some early primary spermatocytes express the *c-kit* gene. Our observations in immunolocalizations confirmed that the kit protein is synthesized in gonocytes of neonates as well as in spermatogonia, and possibly early primary spermatocytes, of adults.

In more recent studies of *c-kit* expression in neonates, we have begun to gather evidence supporting a role for this gene in regulating gonocyte development. For this, we utilized Sertoli cell–gonocyte co-cultures in which cell behavior can be observed with some ease and conditions can be experimentally manipulated. Cells were isolated on the day of birth and cultured for as many as 6 days via procedures developed in our lab some time ago (5), and in situ hybridization and immunolocalization, as described previously, were used to determine *kit* expression and protein synthesis, respectively, in these co-cultures. Our findings indicate (i) that

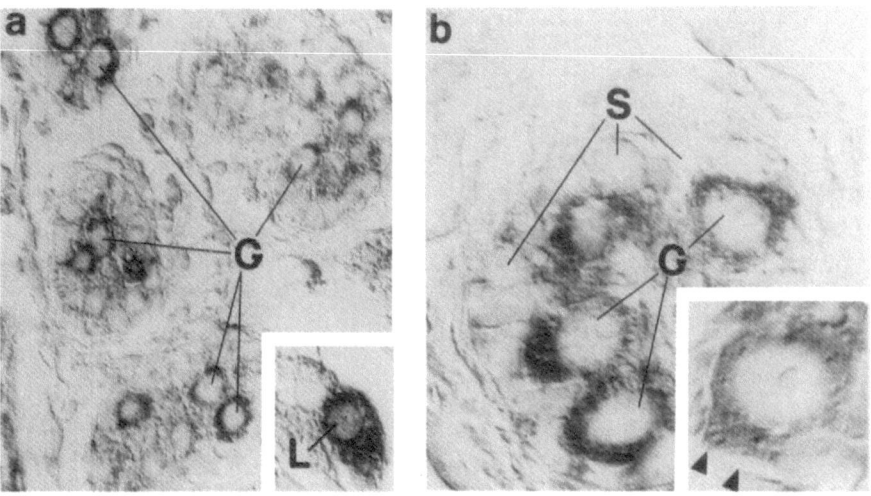

FIGURE 2.2. In situ hybridization carried out on frozen sections of 1-day-old (*a*) and 5-day-old (*b*) testes, with antisense cRNA probes for *c-kit* transcripts. Reaction product in seminiferous cords is restricted to gonocytes (G), some of which have begun to migrate toward the basement membrane (arrowheads, inset in *b*) by day 5. Signal was also seen in apparent Leydig cells (L in inset, *a*) at both ages. Controls exposed to sense probes were uniformly signal negative. Adapted with permission from Orth et al. (15).

c-kit is expressed by gonocytes in co-culture (Fig. 2.3) and that the percentage of *c-kit*-positive cells increases steadily from days 1 through 5 in vitro; (ii) when correlated with morphological evidence of migration, i.e., pseudopod formation, *all* migratory gonocytes express this gene compared to a maximum of about 20% for the gonocyte population as a whole; (iii) inclusion of antibodies against *c-kit* from the day of culture onward results, by day 4, in a steady and significant decrease in the percentage of gonocytes with pseudopods, with apparently no loss of these cells beyond that observed in the absence of antiserum; and (iv) inclusion of recombinant SCF with cultures containing migratory gonocytes causes a transient but significant increase in the percentage of gonocytes with pseudopods (Orth, unpublished findings). In addition, in preliminary immunolocalizations with an antiserum directed against the kit-receptor ligand, SCF, we have detected the presence of this protein on the surfaces of virtually all neonatal Sertoli cells in co-culture (Orth, preliminary observations).

Thus, our observations to date with this in vitro system indicate that neonatal gonocytes express the *c-kit* gene and bear kit receptors and that neonatal Sertoli cells produce SCF. Our results further suggest that interactions between this receptor on gonocytes and its ligand, possibly on Sertoli cell surfaces, are impor-

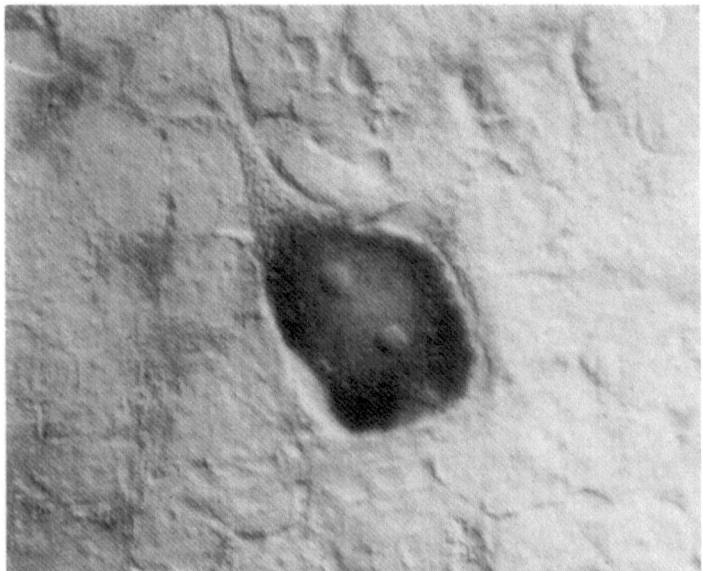

FIGURE 2.3. In situ hybridization carried out on co-cultures prepared on the day of birth and maintained for 5 days in vitro, with antisense cRNA probes specific for *c-kit* transcripts. Reaction product was restricted to gonocytes, as seen in this example of a gonocyte with an obvious pseudopod. Other cells in this view, essentially signal negative, are Sertoli cells. No signal was seen in parallel incubations with sense probes.

tant in maintaining migratory activity of neonatal gonocytes. Our future studies will be aimed at clarifying the precise mechanism whereby these interactions influence gonocyte development.

Summary

In recent years we have come to appreciate that many of the critical events that ensure normal fertility of the adult occur during perinatal development, when the somatic and germ cell populations are differentiating and the basis for spermatogenesis is established. Our recent investigations have underscored the crucial role of cell–cell interactions in these developmental processes and have identified some of the factors involved. Thus far, we know that Sertoli cells adhere to each other and to gonocytes via NCAM expressed on the surfaces of both cell types and that the NCAM isoform expressed is consistent with a role for this molecule in both tight adhesion between cells and in NCAM-based modifications in gene expression. In addition, we have found that neonatal gonocytes express the *c-kit* gene and that neonatal Sertoli cells produce its ligand, stem cell factor, and have accumulated evidence suggesting that this system of cell–cell interaction is vital for migration of gonocytes, at least in vitro. Precise cellular mechanisms involved in these important developmental events remain to be identified and will constitute the focus of our future studies.

References

1. Orth JM, McGuinness MP. Development of postnatal gonocytes in vivo and in vitro. In: Bartke A, ed. Function of somatic cells of the testis. New York: Springer-Verlag, 1994:321–43.
2. Orth JM. Proliferation of Sertoli cells in fetal and postnatal rats: a quantitative autoradiographic study. Anat Rec 1982;203:485–92.
3. McGuinness M, Orth JM. Reinitiation of gonocyte mitosis and movement of gonocytes to the basement membrane in testes of newborn rats in vivo and in vitro. Anat Rec 1992;233:527–37.
4. McGuinness M, Orth JM. Gonocytes of male rats resume migratory activity postnatally. Eur J Cell Biol 1992;59:196–210.
5. Orth JM, Boehm R. Functional coupling of neonatal rat Sertoli cells and gonocytes in co-culture. Endocrinology 1990;127:2812–20.
6. Orth JM, McGuinness M. Neonatal gonocytes co-cultured with Sertoli cells on a laminin-containing matrix resume mitosis and elongate. Endocrinology 1991;129:1119–21.
7. Roosen-Runge EC, Leik J. Gonocyte degeneration in the postnatal male rat. Am J Anat 1968;122:275–300.
8. Orth JM, Jester WF. NCAM mediates adhesion between gonocytes and Sertoli cells in cocultures from testes of neonatal rats. J Androl, 1995;16:389–99.
9. Rutishauser U. Neural cell adhesion molecule and polysialic acid. In: McDonald J, Mecham R, eds. Receptors for extracellular matrix. New York: Academic Press, 1991:131–56.

10. Byers SW, Suijarti S, Jegou B. Cadherins and cadherin-associated molecules in the developing and maturing rat testis. Endocrinology 1994;34:630–4.
11. Kierszenbaum AL. Mammalian spermatogenesis in vivo and in vitro: a partnership of spermatogenic and somatic cell lineages. Endocr Rev 1994;15:116–34.
12. Yoshinaga K, Nishikawa S, Ogawa M, Hayashi S-I, Kunisada T, Fujimoto T, et al. Role of c-kit in mouse spermatogenesis: identification of spermatogonia as a specific site of c-kit expression and function. Development (Camb) 1991;113:689–99.
13. Tajima Y, Sawada K, Morimoto T, Nishimune Y. Switching of mouse spermatogonial proliferation from the c-kit receptor-independent type to the receptor-dependent type during differentiation. J Reprod Fertil 1994;102:117–22.
14. Packer AI, Besmer P, Bacharova RF. Kit ligand mediates survival of Type A and dividing spermatocytes in postnatal mouse testes. Mol Reprod Dev 1995;42:303–10.
15. Orth JM, Jester WF, Qiu J. Gonocytes in testes of neonatal rats express the c-kit gene. Mol Reprod Dev 1996;45:123–31.
16. Dym M, Jia M-C, Dirami G, Price JM, Rabin SJ, Mocchetti I, et al. Expression of c-kit receptor and its autophosphorylation in immature rat Type A spermatogonia. Biol Reprod 1995;52:8–19.

3

Spermatogonial Transplantation

LONNIE D. RUSSELL AND RALPH L. BRINSTER

In the male, gamete production is continuous throughout adult life although some quantitative decline has been observed with advanced age. Only spermatogonia possess the capability of initiating and reinitiating the process of differentiation. In addition, these cells have the capability of self-renewal so that the population of stem cell spermatogonia is not depleted. It is generally thought that only a small percentage of total spermatogonia, the stem cells, possess the capability for self-renewal. Huckins has defined the spermatogonial stem cells morphologically as cells lacking connecting intercellular bridges with other spermatogonia and has called them A_s or $A_{isolated \ cells}$. In other words, the stem cell spermatogonia are isolated from other cells in whole-mount preparations of seminiferous tubules (1). Huckins' scheme for spermatogonial renewal (Fig. 3.1A) is the scheme best accepted by investigators familiar with this field (2–4). Using Huckins' scheme, investigators from de Rooij's laboratory (5) have calculated there are only 35,000 such cells in the testis of an adult mouse.

Another scheme for spermatogonial renewal, developed by Clermont's group, is not favored but has not yet been ruled out. In this scheme, spermatogonial stem cells can arise from cells that have already divided and which, in the process of doing so, have demonstrated some differences in nuclear morphology and have been linked to other spermatogonial cells by intercellular bridges as the consequence of incomplete cytokinesis (Fig. 3.1B). Clermont's group has also indicated that there is a reserve stem cell (A_0) which divides rarely or in conditions when the epithelium has been depleted of other spermatogonia. Because the latter theory of spermatogonial renewal is based on a "dedifferentiation" of cells to renew the stem cell population, there could be potentially manyfold more stem cells in this system as compared with the system proposed by Huckins.

Germ cells are the only cells in the body that transmit their genes to subsequent generations. The germ cell line is, therefore, the most appropriate cell type in

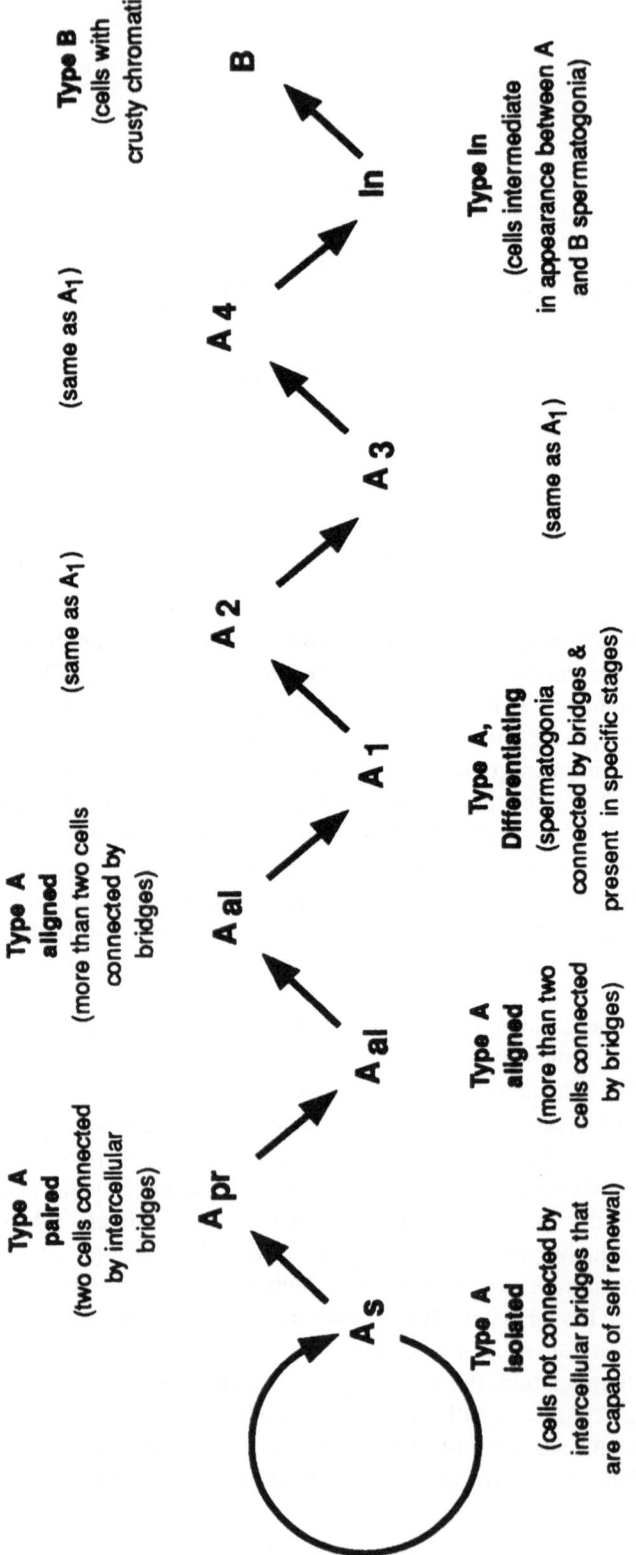

A$_S$ STEM CELL THEORY

A

Type A paired
(two cells connected by intercellular bridges)

Type A aligned
(more than two cells connected by bridges)

(same as A$_1$)

(same as A$_1$)

Type B
(cells with crusty chromatin)

A$_s$

A$_{pr}$

A$_{al}$

A$_{al}$

A$_1$

A$_2$

A$_3$

A$_4$

In

B

Type A isolated
(cells not connected by intercellular bridges that are capable of self renewal)

Type A aligned
(more than two cells connected by bridges)

Type A, Differentiating
(spermatogonia connected by bridges & present in specific stages)

(same as A$_1$)

Type In
(cells intermediate in appearance between A and B spermatogonia)

B

A₀ / A₁ STEM CELL THEORY

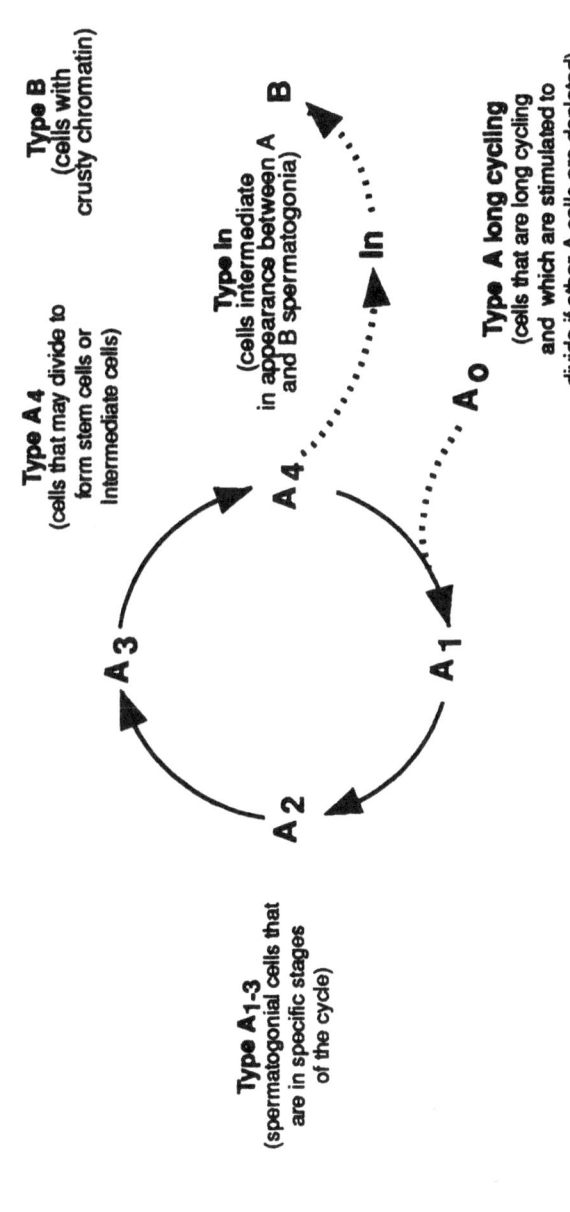

FIGURE 3.1. Two schemes of spermatogonial renewal. (A) The A_s cell is the stem cell, capable of self-renewal or division to form cells that are connected by intercellular bridges (A_{paired} or A_{pr}). The A_{pr} cells divide to form cells aligned in chains connected by intercellular bridges ($A_{aligned}$ or A_{al}), and these in turn divide to form a succession of differentiated stem cells called A_1, A_2, A_3, A_4, In (intermediate), and B (type B). (B) The A_4 cells can either divide to form A_1 cells or divide to form Intermediate (In) spermatogonia. The A_0 cells are suggested to be slow dividing and function in self-renewal or alternatively in forming A_1 cells if the epithelium has been depleted of more advanced types of spermatogonia.

which manipulation of the genome can be undertaken to modify genetic characteristics permanently in a species. In the male, the small number, difficulty in
identification, and confined location of spermatogonial stem cells within the testis
make them relatively inaccessible to genetic manipulation.

Mouse-to-Mouse Germ Cell Transplantation

A breakthrough in technology occurred in Brinster's laboratory when it was
shown that spermatogenesis could be initiated and maintained by intraluminal
transplantation of a mixture of isolated testis cell types from a normal mouse to a
recipient mouse with the same histocompatible background (6, 7). Proof that
transplantations were successful was threefold. First, some of the recipient animals were genetically sterile (W-locus) with a defect affecting germ cell development or had been treated with busulfan to induce temporary loss of all but a few
spermatogonia. Second, the transferred cells bore the *LacZ* gene, the expression of
which was detectable histochemically in germ cells using the substrate X-gal.
Third, offspring were sired from matings of the transplanted animal that bore the
LacZ gene.

The method used for cell transplantation was to microinject a small amount of
germ cell suspension containing dye into the lumen of a seminiferous tubule of the
recipient mouse. The volume injected (20–40 μl) was more than could be handled
by a single seminiferous tubule and thus the injected material entered the rete and
flowed into 70%–90% of surface seminiferous tubules (Fig. 3.2). Thus, most
seminiferous tubules were exposed to the cell suspension. Months later, analysis
of the number of tubules in which spermatogenesis could be found in cross
sections revealed that 31% of the testis contained some degree of spermatogenesis
(8), although the filling efficiency of dye on the surface of the testis and the
spermatogenic efficiency as seen in cross sections of internally positioned tubules
may not be directly comparable.

Because the animals were killed more than 1 year after injection and spermatogenesis was still active, it is presumed that it is the self-renewing stem cell
spermatogonia that had seeded within the seminiferous tubule. Seeding of advanced germ cell types would not have resulted in the presence of spermatogonia
(8) and would not have resulted in continuous spermatogenesis over a long period
of time. Given that intraluminal placement of germ cell suspensions containing
spermatogonia produced active spermatogenesis, it is assumed that spermatogonia were translocated from the apex of Sertoli cells to the base of the seminiferous epithelium (Fig. 3.3). Selective uptake of one germ cell type from a much
larger population of cells within the suspension implies that there are cell-surface
molecules that recognize spermatogonia and initiate cytoskeletal changes in Sertoli cells to transport them basally. In doing so, spermatogonia must move from
the adluminal to the basal compartment, a movement not known to occur naturally
in any adult mammalian species. The movement is similar to that of the neonatal

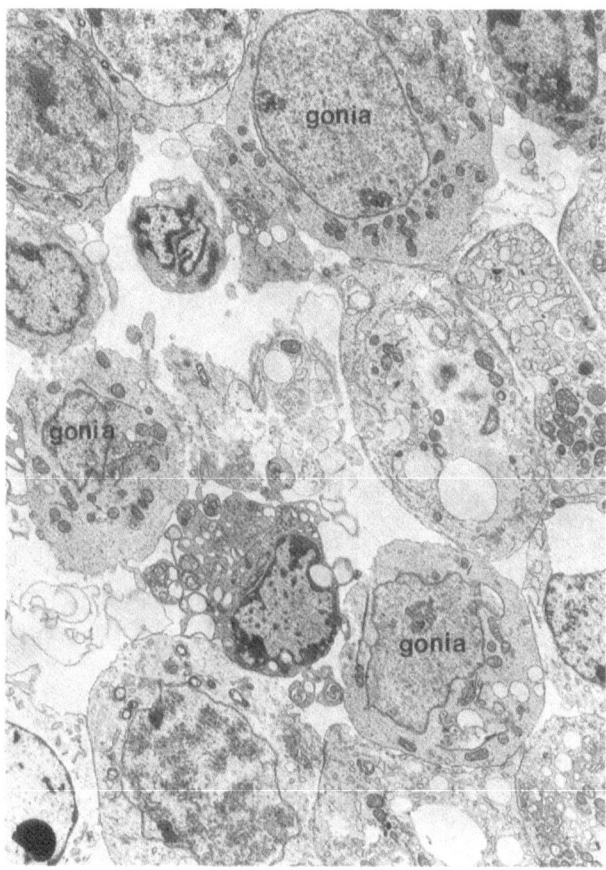

FIGURE 3.2. Electron photomicrograph shows the results of injecting the germ cell suspension of a mouse into a recipient mouse seminiferous tubule. A variety of cell types are shown. Type A spermatogonia are indicated (gonia), but it is not known if these are stem cell spermatogonia.

spermatogonia from the central region of seminiferous tubules to the basal lamina (9), although in the prepubertal animal there is no Sertoli–Sertoli junction to traverse. Alternatively, the movement may occur as an adaptation to conditions that deplete the seminiferous epithelium of germ cells and cause disorganization of the epithelium, and one which can rescue the spermatogenic process with the few stem cell spermatogonia that remain. More work is needed to show how this process occurs.

This study employed transplantation of mouse cells from a donor to a recipient that was genetically sterile. Another recently published study using the rat showed that intraluminal transplantation of cells from one rat to another rat that was fertile

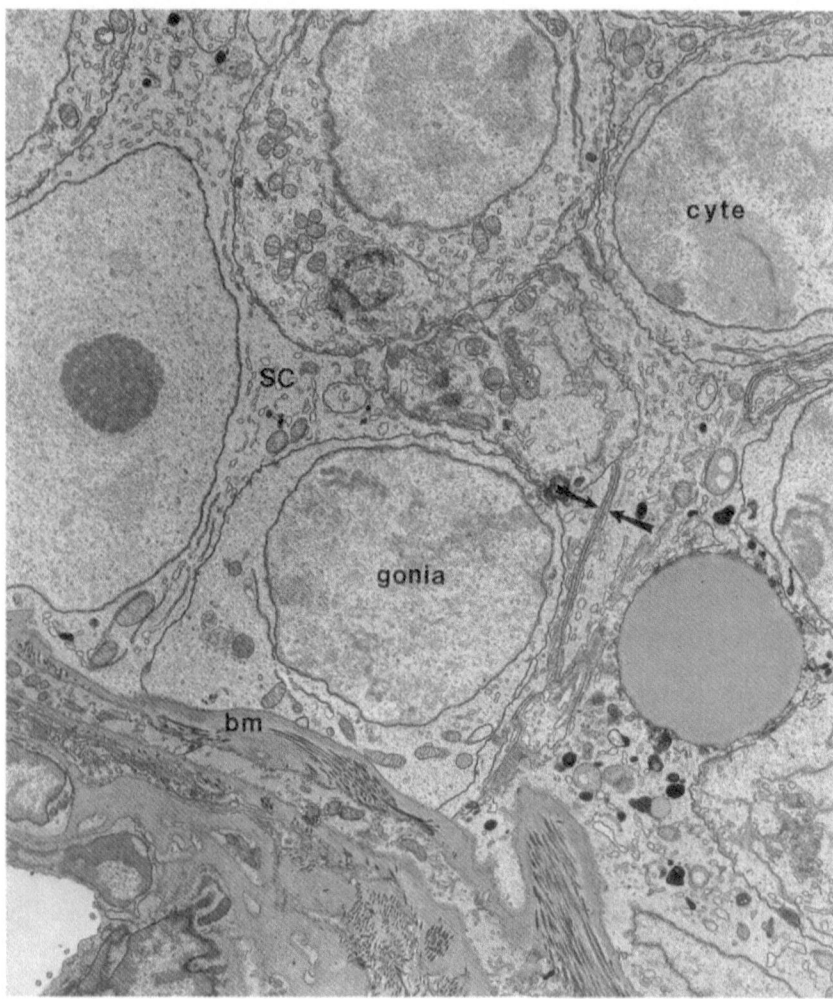

FIGURE 3.3. Mice with the W-locus are normally sterile and may contain rare spermatogonia. The wall of a seminiferous tubule from a W/W genotype mouse having undergone a transplantation with cells from a fertile mouse is shown. A type A spermatogonium (gonia; specific type unknown) has one flattened surface facing the basal membrane (bm) and a rounded surface in contact with surrounding Sertoli cells (SC). The Sertoli–Sertoli junction extends from the spermatogonium upward (opposing arrows). The surrounding Sertoli cell and spermatocytes appear healthy and all are derived from donor cells.

resulted in intraluminal spermatogenesis (10). The authors claimed that all cells of the cell association were present and showed that the intraluminal spermatogenesis was stage-synchronized with the surrounding epithelium of the seminiferous tubule. Spermatogenesis was established with Sertoli cells and peritubular cells, presumably from the injected cell population. It is possible that, under conditions in which spermatogenesis in the recipient is active, the injected spermatogonia are not recognized and taken into the epithelium, but set up intraluminal spermatogenesis in coordination with other injected cell types.

There are several important questions raised by intraspecies transplant protocols as well as potential implication bearing on the direction of future research. One such question is, "Can the efficiency of transplantation be improved if spermatogonia are isolated and purified before transplantation?" At the writing of this chapter, this question remains unanswered. However, we have examined recipient testes at 2 and 7 days after intratubular injection and found all intraluminal germ cells were lacking. By quantification of basal compartment spermatogonia, it was found that their numbers were always greater in transplanted than in nontransplanted tissues, a result suggesting that transport basally of spermatogonia had already occurred. Other cells may have been flushed down the tubular lumen or rapidly degraded by Sertoli cells. The lack of intraluminal cells also points out that there is specificity of Sertoli cell recognition of spermatogonia because cell types other than spermatogonia are not seen at these short intervals after transplantation.

Can spermatogonia be cultured? Techniques for the isolation of type A spermatogonia to about 90% purity have been reported [see (11)]. Organ culture experiments have proven successful for the study of spermatogonial differentiation (12). Van Pelt et al. (13), using the vitamin A-deficient animal to isolate spermatogonia, recently reported that 1×10^5 spermatogonia could be isolated from one testis and that spermatogonial cell preparations were 75%–90% pure. After 3 days of culture the viability of cells was more than 95%. In the vitamin A-deficient animal there appears to be a relative accumulation if not an absolute accumulation of spermatogonia, while virtually all other cell types in the epithelium did not develop during the period of vitamin A deficiency. A method to enhance the purification of spermatogonia, such as using the vitamin A-deficient animal, may allow more ready isolation of these cells to improve the yield. The availability of enriched populations of spermatogonia should facilitate efforts to culture the stem cell.

Can spermatogonia be transfected before being transplanted back to recipient animals? Limited culture and transfection of germ cells have been reported by Cooker et al. (14). The overall goal is to permanently transfect numerous cells simultaneously or to transfect a portion of the cells and to allow them to multiply in culture.

Can the stem cell be isolated and characterized from a mixed population of type A spermatogonia and, if so, can the kinetics of spermatogonial renewal be studied using the transplant model? These questions may now be answerable given that single cells can be selected from many in a culture system and transplanted into seminiferous tubules. Donor animal spermatogonia could be made to express a reporter gene such as LacZ. Single spermatogonial cells could be injected and their kinetics followed in whole-mount sections at various intervals after injection.

Rat-to-Mouse Germ Cell Transplantation

A second type of germ cell transplantation protocol employed by Brinster's laboratory, between two different species, was termed xenogeneic (15). Germ cells isolated from Sprague-Dawley transgenic rats were injected into recipient male mice that were immunodeficient (nude or SCID [severe combined immunodeficiency] mice) whose spermatogenesis was previously suppressed with an injection of busulfan. A reporter (LacZ) transgene that was present in the donor rats was used to identify cells which were successfully transplanted. A second proof of transplantation success was the presence within the epididymis of spermatozoa morphologically characteristic of the rat. Because a low dose of busulfan causes only a temporary suppression of mouse spermatogenesis, the animals examined after 6 months contained evidence of both rat and mouse spermatogenesis. Electron microscopy was also used to determine if spermatogenesis was from rat or mouse stem cells. The shape of the spermatid head, the degree of condensation of the spermatid head, and the positioning of organelles within germ cells provided unequivocal identification of the species origin of cell types (16).

Although intermixing of mouse and rat spermatogenesis occurred in xenogeneic transplants, there was no apparent intermixing of cell types within a given region of the tubule. Thus, spermatogonia that were seen at the base of the tubule were assumed to be from the species whose cell types were seen within the same tubule, and spermatogonia were always seen at the base of the tubule resting on the basal lamina and below Sertoli–Sertoli junctions. Not only can Sertoli cells apparently recognize spermatogonial cells of the same species and translocate them from the site of injection to their normal position, but they can also recognize them from another species.

As rat spermatogonia transplanted into mice differentiate, they form cell associations characteristic of the rat. It is not known if the timing of the cell kinetics in the transplanted rat cells is similar to that of the mouse or of the rat. However, to form the cell associations characteristic of the rat the kinetics appear to be regular, indicating that rat spermatogonia can divide regularly to form cell associations characteristic of the rat even though they exist within an environment normally supporting mouse cells.

Summary

Spermatogenesis can be initiated and spermatozoa formed by transplantation of a mixture of testis cells from donors into the seminiferous tubules of recipient mice that either are genetically sterile or have had most of their germ cell population depleted by busulfan treatment. Spermatogenesis can also be initiated and completed by spermatogonial transplantation from a mixture of rat donor testis cells to immunodeficient mice. Stem cell spermatogonia must be selectively recognized and must be transported to the basal compartment to initiate and maintain spermatogenesis. These experiments are important steps in studying spermatogonial kinetics, isolation of stem cells, culturing spermatogonial cells, and also modification of the genome before transplantation into a recipient.

Acknowledgments. Financial support for this work was in part from the National Institutes of Health (NICHD23657), USDA/NRI Competitive Grants Program (95–37205–2353), and the Robert J. Kleberg, Jr. and Helen C. Kleberg Foundation.

References

1. Huckins C. Spermatogonial intercellular bridges in whole-mounted seminiferous tubules from normal and irradiated rodent testis. Am J Anat 1978;153:97–122.
2. Oakberg EF, Huckins C. Spermatogonial stem renewal in the mouse as revealed by ^{3}H-thymidine labeling and irradiation. In: Cairnnie AB, Lala PK, Osmond DG, eds. Stem cells and renewing cell populations. New York: Academic Press, 1976:287–302.
3. Huckins C, Oakberg EF. Morphological and quantitative analysis of spermatogonia in mouse testis using whole mounted seminiferous tubules. II. The irradiated testis. Anat Rec 1978;192:529–54.
4. de Rooij DG, van Dissel Emiliani FM, van Pelt AM. Regulation of spermatogonial proliferation. Ann N Y Acad Sci 1989;564:140–53.
5. Tegelenbosch RAJ, De Rooij DG. A quantitative study of spermatogonial multiplication and stem cell renewal in the C3H/101 F1 hybrid mouse. Mutat Res 1993;290:193–201.
6. Brinster RL, Avarbock MR. Germline transmission of donor haplotype following spermatogonial transplantation. Proc Natl Acad Sci USA 1994;91:11303–7.
7. Brinster RL, Zimmerman, JW. Spermatogenesis following male germ-cell transplantation. Proc Natl Acad Sci USA 1994;91:11298–302.
8. Russell LD, França LR, Brinster RL. Ultrastructural observations of spermatogenesis in mice resulting from transplantation of mouse spermatogonia. J Androl 1996;17:603–14.
9. McGuinness MP, Orth JM. Gonocytes of male rats resume migratory activity postnatally. Eur J Cell Biol 1992;59:196–210.
10. Jiang F-X, Short RV. Male germ cell transplantation in rats: apparent synchronization of spermatogenesis between host and donor seminiferous epithelia. Int J Androl 1995;18:326–30.
11. Dym M. Spermatogonial stem cells of the testis. Proc Natl Acad Sci USA 1994;91:11287–9.
12. Boitani C, Giuditta P, Tiziana M. Spermatogonial cell proliferation in organ culture of immature rat testis. Biol Reprod 1993;48:761–7.
13. van Pelt AMM, Morena AR, van Dissel-Emiliani FMF, Boitani C, Gaemers IC et al. Isolation of the synchronized A spermatogonia from adult vitamin A-deficient rat testes. Biol Reprod 1996;55:439–44.
14. Cooker LA, Brooke CD, Kumari M, Hofman M, Millan J, Goldberg E. Genomic structure and promoter activity of the human testis lactate dehydrogenase gene. Biol Reprod 1993;48:1309–19.
15. Clouthier DE, Avarbock MR, Maika SD, Hammer RE, Brinster RL. Rat spermatogenesis in mouse testes following spermatogonial stem cell transplantation. Nature (Lond) 1996;381:418–21.
16. Russell LD, Brinster RL. Ultrastructural observations of spermatogenesis following transplantation of rat testis cells into mouse seminiferous tubules. J Androl 1996;17:615–27.

4

Telomerase Activity Is Lost During Male Germ Cell Differentiation

Martin Dym and Neelakanta Ravindranath

During embryonic development, male primordial germinal cells, which arise from the yolk sac, migrate to the bilateral gonadal ridges and proliferate (1). They interact with coelomic epithelial and mesenchymal cells to organize into testicular cords. The primordial germinal cells become gonocytes, the precursor cells of spermatogonia. The epithelial and mesenchymal cells derived from the mesonephros give rise to Sertoli cells, peritubular myoid cells, and interstitial cells of the testis (2). At birth, the testicular cords display gonocytes and undifferentiated Sertoli cells (3). Gonocytes proliferate and form type A spermatogonia during the early neonatal period (4). The unique process of morphological and functional differentiation of type A spermatogonia into spermatozoa is termed spermatogenesis. Unlike many other cells in the body that only undergo functional differentiation, type A spermatogonia also undergo morphological differentiation. The important features of type A spermatogonia are (i) they can proliferate; (ii) they can renew and self-maintain their numbers; (iii) they can produce a large number of functional progeny, i.e., the differentiated germ cells; and (iv) they can regenerate the advanced germ cell types after injury. These features are the hallmarks of "stem cells" (5). Thus, type A spermatogonia could be considered as the "stem cells of the germ cell lineage" (6).

The molecular basis for self-renewal, differentiation, and death of spermatogonial stem cells has not been elucidated. Both extrinsic and intrinsic factors are known to regulate stem cells (7). Although the pituitary hormones, follicle-stimulating hormone (FSH) and luteinizing hormone (LH), could be considered as extrinsic factors, they have no direct effect on the spermatogonial stem cells because receptors for FSH and LH are absent on the surface of these stem cells. FSH binds to specific cell-surface receptors on Sertoli cells in the seminiferous tubular compartment whereas LH binds exclusively to receptors on the surface of Leydig cells in the interstitial compartment (8, 9). Sertoli cells also possess receptors for androgens (9). In the absence of receptors for FSH, LH, and androgens in the spermatogonial stem cells, it could be presumed that their renewal, differentiation, and death are mediated by factors released by Sertoli cells in response to FSH and androgens. An example of one such factor released by Sertoli cells in response to FSH is stem cell factor (10). The receptor for stem cell factor, the *c-kit* receptor, is expressed in the type A spermatogonial stem cells (11). The biological role of the stem cell factor/*c-kit*-receptor system in type A spermatogonial renewal and differentiation is still not known.

Intrinsic factors that may regulate the capacity of spermatogonial stem cells to renew or differentiate are constraints of space, adhesive interactions with the basement membrane, and other autonomous mechanisms involved in stem cell proliferation and renewal. One autonomous mechanism that has gained much attention in recent years is the length of the telomeres which cap the ends of chromosomes. It was suggested that linear genomic DNA loses terminal sequences during each replication of the cell (12, 13). The human telomeric DNA sequences have been identified to be repeats of TTAGGG (14). Normal human somatic cells lose terminal repeats of TTAGGG at a rate of 50–200 bp per population doubling (15). In contrast, spermatozoa, which are formed after several mitotic divisions of spermatogonia and two meiotic divisions of spermatocytes, exhibit long tracts of repetitive telomeric DNA (16), suggesting that there must be a mechanism by which the telomeric DNA loss can be prevented. Telomere shortening has also not been observed in many types of cancer cells (17). Telomerase, a reverse transcriptase enzyme containing an RNA template for telomeric sequence, synthesizes telomeres de novo and specifically adds them to the chromosome 3'-ends (18). A sensitive polymerase chain reaction- (PCR-) based assay has been developed to measure telomerase activity in cells and tissues (19). Using this assay, we investigated the expression of telomerase activity in the developing rat testis and in isolated germ cell types from immature and adult rats.

Materials and Methods

Isolation and Characterization of Type A Spermatogonia from Immature Rats

Type A spermatogonia from 9-day-old rats were isolated by the method of sedimentation velocity at unit gravity (11). The testes were excised and decapsulated. The decapsulated testes were subjected to collagenase (1.5 mg/ml) and DNAse (1 μg/ml) for 15 min at 34°C. Seminiferous cord fragments obtained were washed thoroughly to remove all the interstitial cells and incubated with collagenase (1.5 mg/ml), hyaluronidase (1.5 mg/ml), and DNAse (1 μg/ml) for 20–30 min at 34°C. The dispersed cells were washed and filtered through 80-μm and 40-μm nylon mesh, successively. The single cells in suspension were separated by sedimentation velocity at unit gravity at 4°C, with use of a 2%–4% bovine serum albumin (BSA) gradient. The cells were allowed to sediment for a standard period of 2.5 h, and then 35 fractions of 15 ml each were collected at 90-s intervals. The cells in each fraction were examined under a phase-contrast microscope, and fractions containing cells of similar morphology and size were pooled and spun down by low-speed centrifugation and resuspended in appropriate medium.

The cells in the enriched spermatogonial fractions were approximately 20–25 µm in diameter and had large spherical nuclei containing several prominent nucleoli. The cytoplasm was characterized by organelles located mostly in the perinuclear region. The pooled spermatogonial fractions were subjected to differential plating in medium containing 5% fetal bovine serum for 4 h at 34°C to eliminate the contaminating cells. The freshly isolated spermatogonial cells and the enriched population recovered after differential plating were centrifuged at $90 \times g$ for 5 min onto glass slides in a Cytospin centrifuge. The cells were fixed and permeabilized with ice-cold methanol for 5 min and subjected to streptavidin-biotin peroxidase immunostaining with Histostain-SP kits (Zymed Laboratories, Burlingame, CA) using a rabbit polyclonal antibody for the mouse *c-kit* receptor as the primary antibody (11). The spermatogonial cells exhibited orange-red immunostaining for the *c-kit* receptor, but the contaminant cells were devoid of any staining. Approximately 1000 cells were counted in several different fields, and the percentages of stained and unstained cells were established. The purity of the freshly isolated spermatogonia was in the range of 80–85%. After removal of Sertoli cell and peritubular myoid cell contamination by differential plating, the purity of the population was in the range of 95–100%.

Isolation of Pachytene Spermatocytes and Round Spermatids from Adult Rat Testis

Using the same technique of sedimentation velocity at unit gravity, the pachytene spermatocytes and round spermatids were isolated from the adult rat testis. Based on a morphological approach, the percentage of purity of pachytene spermatocytes and round spermatids in the enriched populations was established. While the purity of pachytene spermatocytes ranged from 80% to 90%, the purity of round spermatids was between 70% and 75%. The percentage of contamination with spermatogonia was much higher in the round spermatid population.

Isolation of Spermatozoa from the Rat Epididymis

Spermatozoa were mechanically expressed from the epididymis of adult rats.

Telomerase Assay

Telomerase activity in the lysed homogenates of the developing testis and isolated germ cell types was measured using the TRAP-eze Telomerase Detection Kit (Oncor, Gaithersburg, MD). Briefly, the TS primer provided in the kit was radiolabeled with γ-^{32}P-ATP using T4 polynucleotide kinase. A reaction mixture consisting of labeled TS primer, tartrate-resistant acid phosphatase (TRAP) primer mix, deoxynucleoside triphosphatases (dNTPs), Taq polymerase, and cell or tissue extract was incubated at 30°C for 10 min. PCR amplification was performed at 94°C for 30 s followed by 60°C for 30 s for 25–30 cycles. Appropriate positive, negative, and PCR controls were included in the assay. The reaction mixture was run on a 12.5% nondenaturing polyacrylamide gel in 0.5× TBE (Tris-borate + EDTA) buffer for 1 h. The gel was dried, and assay products were visualized by autoradiography.

Results and Discussion

In the developing rat testis, telomerase activity was expressed at all ages from birth to adulthood. However, the telomerase activity appeared to increase from birth until week 4 after birth and then declined progressively to low levels by the tenth week of age. Contrary to our findings, Prowse and Greider (20) observed telomerase activity in the testis of 6- to 16-week-old mice, but not in the testis of 4-week-old mice. The activity seen was attributed to the expression of telomerase in the pachytene spermatocytes. The spermatogonial stem cell population, which is more enriched in the testis during the first few weeks after birth, did not appear to express high levels of activity or the assay used by Prowse and Greider (20) was not sensitive enough to detect the presence of activity. To clarify the expression of telomerase activity in spermatogonial stem cells, we isolated these stem cells from the immature rat and analyzed them for telomerase activity. In agreement with the expression of telomerase activity in hematopoietic stem cells in the body (21–23), spermatogonial stem cells from 9-day-old rats expressed very high levels of activity. To determine at which point in the spermatogonial differentiation pathway the telomerase activity is lost, the extracts of enriched populations of pachytene spermatocytes, round spermatids, and epididymal spermatozoa were assayed for telomerase activity. Both the enriched populations of pachytene spermatocytes and round spermatids expressed telomerase activity, although much less intensely than in spermatogonial stem cells. The activity observed may be the result of definite expression of telomerase in these populations or contaminating spermatogonial cells. This possibility needs to be investigated further. Interestingly, epididymal spermatozoa did not exhibit any telomerase activity in agreement with the results in mature human sperm (24).

Summary

In summary, spermatogonial stem cells express high levels of telomerase activity, and this activity declines progressively through the stages of pachytene spermatocytes and round spermatids as germ cell differentiation proceeds. A total loss of telomerase activity is observed in epididymal spermatozoa. Although germ cell differentiation leads to a loss of telomerase activity, it is not yet known precisely at which step of germ cell development the telomerase activity is lost.

References

1. Tam PPL, Snow MHL. Proliferation and migration of primordial germinal cells during compensatory growth in mouse embryos. J Embryol Exp Morphol 1997;64:133–47.
2. Pelliniemi LJ, Frojdman K, Paranko J. Embryological and prenatal development and function of Sertoli cells. In: Russell LD, Griswold MD, eds. The Sertoli cell. Clearwater, FL: Cache River Press, 1993:87–114.
3. Clermont Y, Perey B. Quantitative study of the cell population of the seminiferous tubules in immature rats. Am J Anat 1957;100:241–67.
4. McGuinness MP, Orth JM. Reinitiation of gonocyte mitosis and movement of gonocytes to the basement membrane in testes of newborn rats in vivo and in vitro. Anat Rec 1992;233:527–37.
5. Potten CS, Loeffler M. Stem cells: attributes, cycles, spirals, pitfalls and uncertainties. Lessons for and from the crypt. Development (Camb) 1990;110:1001–20.
6. Dym M. Spermatogonial stem cells of the testis. Proc Natl Acad Sci USA 1994; 91:11287–9.
7. Morrison SJ, Shah NM, Anderson DJ. Regulatory mechanisms in stem cell biology. Cell 1997;88:287–98.
8. Griswold MD. Action of FSH on mammalian Sertoli cells. In: Russell LD, Griswold MD, eds. The Sertoli cell. Clearwater, FL: Cache River Press, 1993:493–508.
9. Sar M, Hall SH, Wilson EM, French FS. Androgen regulation of Sertoli cells. In: Russell LD, Griswold MD, eds. The Sertoli cell. Clearwater, FL: Cache River Press, 1993:509–16.
10. Rossi P, Dolci S, Albanesi C, Grimaldi P, Ricca R, Geremia R. Follicle-stimulating hormone induction of steel factor (SLF) mRNA in mouse Sertoli cells and stimulation of DNA synthesis in spermatogonia by soluble SLF. Dev Biol 1993;155:68–74.
11. Dym M, Jia MC, Dirami G, et al. Expression of c-kit receptor and its phosphorylation in immature rat type A spermatogonia. Biol Reprod 1995;52:8–19.
12. Watson JD. Origin of concatemeric T7 DNA. Nat New Biol 1972;239:197–201.
13. Olovnikov AM. A theory of marginotomy. The incomplete copying of template margin in enzymic synthesis of polynucleotides and biological significance of the phenomenon. J Theor Biol 1973;41:181–90.
14. Moyzis RK, Buckingham JM, Cram LS, et al. A highly conserved repetitive DNA sequence, $(TTAGGG)_n$, present at the telomeres of human chromosomes. Proc Natl Acad Sci USA 1988;85:6622–6.
15. Harley CB, Futcher AB, Greider CW. Telomeres shorten during ageing of human fibroblasts. Nature (Lond) 1990;345:458–62.
16. Hastie ND, Dempster M, Dunlop MG, Thompson AM, Green DK, Allshire RC. Telomere reduction in human colorectal carcinoma and with ageing. Nature (Lond) 1990;346:866–8.
17. Counter CM, Avilion AA, Le Feuvre CE, et al. Telomere shortening associated with chromosome instability is arrested in immortal cells which express telomerase activity. EMBO J 1992;11:1921–9.
18. Greider CW, Blackburn EH. Identification of a specific telomere terminal transferase activity in Tetrahymena extracts. Cell 1985;43:405–13.
19. Kim NW, Piatyszek MA, Prowse KR, et al. Specific association of human telomerase activity with immortal cells and cancer. Science 1994;266:2011–5.
20. Prowse KR, Greider CW. Developmental and tissue-specific regulation of mouse telomerase and telomere length. Proc Natl Acad Sci USA 1995;92:4818–22.

21. Hiyama K, Hirai Y, Kyoizumi S, et al. Activation of telomerase on human lymphocytes and hematopoietic progenitor cells. J Immunol 1995;155:3711–5.
22. Broccoli D, Young JW, de Lange T. Telomerase activity in normal and malignant hematopoietic cells. Proc Natl Acad Sci USA 1995;92:9082–6.
23. Chiu CP, Dragowska W, Kim NW, et al. Differential expression of telomerase activity in hematopoietic progenitors from adult human bone marrow. Stem Cells (Basel) 1996;14:239–48.
24. Wright WE, Piatyszek MA, Rainey WE, Byrd W, Shay JW. Telomerase activity in human germline and embryonic tissues and cells. Dev Genet 1996;18:173–9.

Part II

Meiosis

5

Regulation of Sister-Chromatid Cohesion During *Drosophila* Meiosis

SHARON E. BICKEL AND TERRY L. ORR-WEAVER

When cells divide it is imperative that each daughter cell receive the correct complement of chromosomes. Missegregation of chromosomes leads to aneuploidy, which results in catastrophic consequences for both meiotic and mitotic cells. Much current research focuses on cell-cycle controls that elicit the metaphase/anaphase transition as well as the molecules that are necessary for the proper movement of chromosomes during anaphase. However, for chromosome segregation to proceed accurately there must be a mechanism to keep sister chromatids attached until anaphase. Only if sister chromatids remain connected to each other are they able to form stable bipolar attachments to spindle microtubules. Therefore, sister-chromatid cohesion prevents individual chromatids from segregating randomly to either pole and ensures that chromosomes are partitioned correctly (Fig. 5.1).

Differential Regulation of Mitotic and Meiotic Cohesion

Cytological examination of chromosome behavior has revealed striking differences in the manner in which cohesion is regulated during meiotic and mitotic divisions [for review, see (1)]. During mitosis, sister chromatids are held together along their entire length. At the metaphase/anaphase transition all cohesion is abolished, allowing sisters to separate from each other and move to opposite poles (Fig. 5.1).

The modulation of cohesion during meiosis is regulated in a more complex manner (Fig. 5.2). In meiosis I, sister chromatids must remain together as a unit when the homologues segregate to opposite poles. Sister chromatids maintain connections along their entire length until anaphase I, when attachment between the arms is destroyed. Presumably, ablation of arm cohesion is required to allow

NORMAL SEGREGATION IN MITOSIS

cohesion

↓

orientation

↓

attachment

↓

tension

↓

segregation

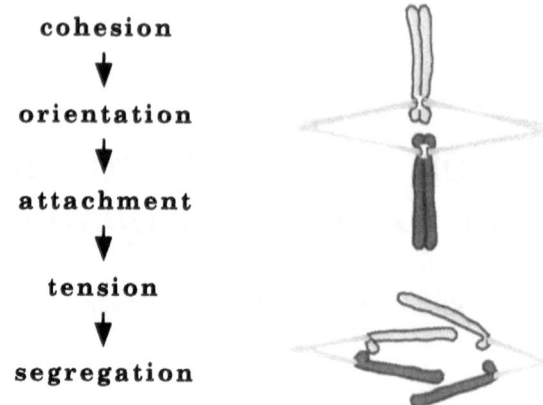

CONSEQUENCES OF NO COHESION

random orientation and
segregation of chromatids

↓

aneuploidy

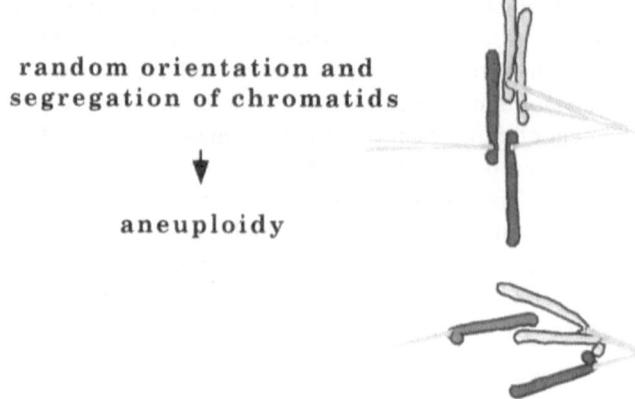

FIGURE 5.1. During cell division, sister-chromatid cohesion is required for chromosome segregation to proceed normally (top). Without cohesion, sister chromatids are not constrained to attach to microtubules emanating from opposite poles. Individual chromatids are able to segregate randomly resulting in aneuploidy (bottom).

the separation of homologues that have undergone reciprocal exchange. After arm cohesion is released at anaphase I, sister chromatids remain connected at their centromeric regions. This association is essential for sisters to align properly on the metaphase II plate (2, 3). At the second metaphase/anaphase transition, centromeric cohesion is lost, and sisters move to opposite poles.

Although cohesion is essential during meiosis and mitosis, it remains unknown whether the molecules controlling cohesion are conserved between these two types of division. Moreover, the stepwise dissolution of cohesion during meiosis

MEIOSIS I

**homolog association
recombination**

**bipolar orientation
of homologs**

loss of arm cohesion

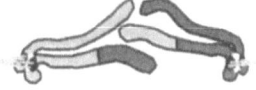

segregation of homologs

MEIOSIS II

**kinetochores face
opposite directions**

**centromeric cohesion
is lost**

sisters segregate

FIGURE 5.2. Cohesion during meiosis is lost in a stepwise manner. Sister chromatids are held together along their entire length until the first metaphase/anaphase transition when arm cohesion is abolished. During meiosis II, cohesion is confined to the centromeric region of the chromosomes. At anaphase II, centromeric cohesion is lost, allowing the sister chromatids to segregate to opposite poles. Only one set of sisters is depicted for the second division.

suggests that the mechanical basis of arm and centromeric cohesion may be different.

Identification of the Proteins That Control Cohesion

Although cohesion is essential, very little is known about its physical nature or its regulation during the cell cycle. Catenation resulting from DNA replication is one potential mechanism for keeping sister chromatids connected (4). A protein required for cohesion could act in a regulatory manner to inhibit topoisomerase II from resolving interlocked helices until anaphase. Another possibility is that a protein or protein complex might promote sister-chromatid cohesion by acting as a structural "glue." Immunolocalization studies have identified several candidate cohesion proteins that localize on chromosomes at points of contact between sisters (5–9). However, although this pattern is consistent with a role in cohesion, the function of such proteins remains to be demonstrated.

A powerful approach to dissecting the regulation of sister-chromatid cohesion is to identify mutations that disrupt cohesion. *Drosophila* serves as an ideal model system because it offers the advantages of genetic, cytological, and molecular analysis. Genes controlling meiotic chromosome segregation are easily isolated on the basis of their mutant phenotype because there are only four chromosomes, and the organism is viable when two of them are monosomic. Thus, mutations that disrupt cohesion and randomize chromosome segregation still yield viable progeny. In addition, a variety of visible markers make it possible to follow the segregation of specific chromosomes genetically. Genetic screens also may be performed to identify interacting proteins either as genetic suppressors or as noncomplementers of an unlinked mutation.

By combining cytological examination of the chromosomes with genetic analysis, one may gain information about the nature of the segregation defect as well as when it first becomes manifest. Sister-chromatid cohesion may be monitored easily in *Drosophila* spermatocytes undergoing meiosis by preparing orcein-stained squash preparations of adult testes (3). In addition, analysis of chromosome behavior during female meiosis in *Drosophila* has recently become accessible through the use of confocal imaging techniques combined with an in vitro activation protocol (10, 11).

Mutations in two *Drosophila* genes, *ord* (orientation disruptor) and *mei-S332*, disrupt meiotic chromosome segregation in both males and females (2, 3, 12–14). Moreover, *ord* and *mei-S332* display cohesion abnormalities at different times, suggesting that the mode of action of their gene products may be distinct. *ord* function is essential early in meiosis I to ensure proper cohesion. In contrast, *mei-S332* becomes necessary after the first metaphase/anaphase transition. Furthermore, recent localization of a MEI-S332-GFP fusion protein has demonstrated that MEI-S332 acts at the centromeric region of meiotic chromosomes to maintain cohesion (15).

The ORD Protein Is Essential for Meiotic Cohesion

The original *ord* mutation was identified because it disrupted meiotic chromosome segregation in both males and females (12). Several additional *ord* alleles have been isolated that exhibit the same behavior (14) (Bickel and Orr-Weaver, unpublished observations). Defects in cohesion are visible in *ord* spermatocytes as early as prometaphase I, when sisters should be attached along their entire length (3, 14). A lack of cohesion during meiosis I would allow individual chromatids to segregate randomly during both meiotic divisions. Indeed, in chromosome segregation assays using marked chromosomes, *ord* mutants undergo missegregation during meiosis I and II.

The frequency of meiotic reciprocal recombination also is severely reduced in *ord* females. However, missegregation in *ord* mutants cannot be explained solely by recombination defects, because exchange chromosomes missegregate as frequently as nonrecombinant chromosomes in *ord* females (12). Also, wild-type *Drosophila* males do not undergo meiotic recombination, but cohesion defects are still manifest in *ord* spermatocytes. These data suggest that the principal function of ORD protein is to establish or maintain meiotic sister-chromatid cohesion. Determining the molecular details of how ORD promotes cohesion will provide valuable insights into this essential aspect of cell division.

Dissecting the Functional Domains of ORD Using Genetic and Molecular Analysis

As the first step toward understanding how the ORD protein controls cohesion, the *ord* gene was cloned, and the molecular lesions associated with several mutations were identified (16). ORD is a novel 55-kDa polypeptide. The C-terminal half of the protein is essential for function. Several mutations that affect both cohesion and recombination map to this region of the protein (Fig. 5.3). In addition, the N-terminus of ORD contains a PEST sequence that may be instrumental in rapid degradation. It is possible that the release of cohesion at the metaphase/anaphase transition is modulated in part by proteolysis of ORD.

Molecular characterization of *ord* mutations was crucial in understanding an unusual genetic phenomenon displayed by specific combinations of mutant *ord* alleles (16). Although they are recessive when in trans to wild-type *ord,* certain missense *ord* alleles are able to poison the residual activity of weaker alleles. In other words, in transheterozygotes, an *ord* mutation with a single amino acid change causes a more severe phenotype than a deficiency that removes the gene.

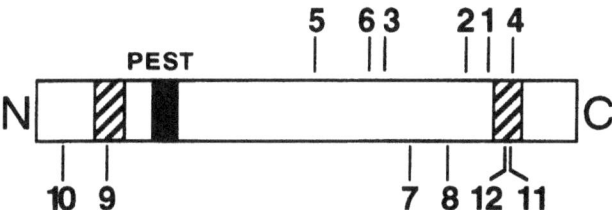

FIGURE 5.3. The relative site of each *ord* mutation is depicted within a schematic representation of the ORD protein. Two hydrophobic intervals are marked by diagonal lines. The PEST motif is shown in black.

This behavior, termed negative complementation, has been observed for genes whose protein products are known to require protein interactions for activity (17–20). All *ord* mutations that exhibit negative complementation are missense changes located within the C-terminal half of the protein (16) (Bickel and Orr-Weaver, unpublished observations). This suggests that the C-terminal part of ORD participates in protein–protein interactions that are necessary for ORD function.

The negative complementation of several *ord* alleles coupled with their molecular analysis has led to our proposal that two activities, protein-binding and cohesion ability, are required for ORD to be fully functional (16). These may be separably mutated (Fig. 5.4A). In our model, ORD must interact with a protein partner to activate its ability to promote cohesion. We hypothesize that a mutation such as *ord⁴* diminishes protein-binding efficiency. However, once binding has taken place, ORD⁴ is able to promote cohesion. This capability would account for the observation that chromosome segregation is almost completely normal in *ord⁴/ord⁴* or *ord⁴/Df* flies (16). In contrast, we suggest that the *ord¹* mutation allows normal protein–protein interactions but disrupts the "active site" required for cohesion activity once binding has taken place. The fact that *ord¹* is recessive indicates that binding affinity of ORD¹ is not greater than wild-type ORD. However, because ORD⁴ binds less well, ORD¹ is able to poison the cohesive ability of ORD⁴ by competing for the protein partner.

Identification of the protein-binding partner required for ORD activity is crucial in understanding the mechanical basis of cohesion. One possibility is that ORD molecules complex with each other. However, our observations indicate that in the yeast two-hybrid assay ORD does not interact with itself (Young et al., unpublished observations). Therefore, we favor the hypothesis that a protein partner yet to be identified is required to interact with ORD to elicit cohesion.

Characterization of New *ord* Alleles

We recently performed a genetic screen to isolate additional *ord* alleles that failed to complement the *ord¹* mutation in males. A null allele, *ord¹⁰*, was isolated that contains a stop codon after 23 amino acids. Isolation of *ord¹⁰* has allowed us to determine the consequences of ablating *ord* activity, and these results are described elsewhere. New missense alleles also were recovered, and these have provided additional information about functional domains that are critical for ORD activity. Interestingly, two missense mutations were isolated that exhibited phenotypes similar to the null allele. One of these, *ord⁹*, is the first missense mutation identified within the N-terminal part of the ORD protein. Our tests indicate that the *ord⁹* allele disrupts both cohesion and recombination.

Another strong missense mutation, *ord¹²*, is located in the C-terminal region of the protein where several other *ord* mutations reside. *ord¹²/ord¹⁰* flies display levels of missegregation that are comparable to flies containing no ORD activity (Tables 5.1 and 5.2). The *ord¹²* mutation converts Val⁴¹⁶ to Asp and lies within an

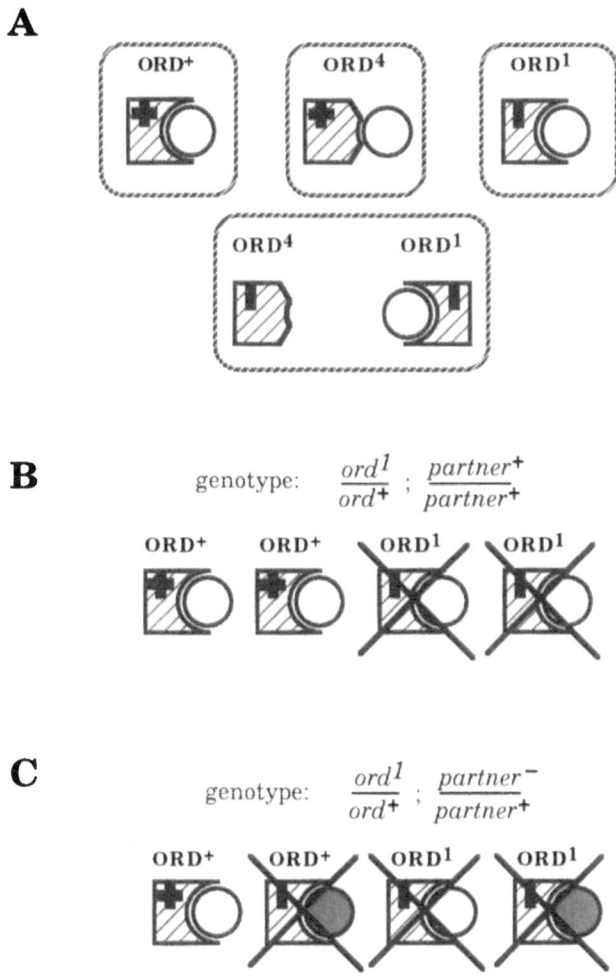

FIGURE 5.4. (*A*) A model explaining the negative complementation of *ord¹* and *ord⁴*. The ORD protein is represented as a hatched polygon and the proposed protein partner as an open circle. When ORD⁺ binds the protein partner, its "active site" (solid black bar) is able to promote cohesion (black cross). The ORD⁴ mutation decreases binding affinity, but once the partner is bound, ORD⁴ protein is functional. The "active site" is mutated in ORD¹, so that even when binding does occur, it cannot promote cohesion. In an *ord⁴/ord¹* fly, ORD¹ competes for the protein partner and prevents ORD⁴ from becoming active. (*B*) *ord¹* is recessive to wild-type *ord*. In an *ord¹/ord⁺* fly, enough functional complex is formed so that cohesion is normal. (*C*) An example of second-site noncomplementation is shown in which an aberrant complex is formed in the presence of a second-site mutation. A defective protein partner is represented as a dark circle. In this case, the protein partner is able to bind to ORD but does not activate ORD's cohesion function. Although wild-type ORD and wild-type partner are still present, the amount of functional complex formed is less than in *B*. If this level is below the threshold required for cohesion, missegregation will occur.

TABLE 5.1. Sex chromosome missegregation in *ord* males.

Genotype	Regular sperm (%)		Exceptional sperm (%)				Total progeny	Total % missegregation
	X	$Y(Y)$	O	$XY(Y)$	XX	$XXY(Y)$		
ord[12]/ord[10]	30.6	21.7	32.7	11.1	3.7	0.3	1810	47.8
ord[12]/ord[8]	53.3	36.9	5.8	2.7	1.3	0.0	4093	9.8
ord[11]/ord[8]	49.3	38.6	7.9	2.6	1.6	0.1	1831	12.2

[a]$y/y + Y$ males were crossed to attached-X, $y2\ su(w^a)\ w^a$ females.

interval that is rich in hydrophobic amino acids. In contrast to *ord[12]*, mutations located nearby at positions 423 (Ser to Arg) and 424 (Ala to Val) are significantly weaker. It is possible that these alleles have milder effects on segregation because their missense amino acid changes do not disrupt the hydrophobic character of the interval so severely as that of *ord[12]*.

Although it severely disrupts ORD function, *ord[12]* exhibits minimal negative complementation. This is in contrast to other missense mutations in the C-terminal part of the protein that do show negative complementation. In males, *ord[12]* weakly poisons the activity of *ord[8]*, and in females it does not appear to affect the function of *ord[8]* at all (Tables 5.1 and 5.2). The total level of missegregation in *ord[8]/Df* males and females is 4.1% and 10.9%, respectively. These results suggest that the binding activity of *ord[12]* is severely compromised by the Val[416] to Asp mutation. This would account for the inability of ORD[12] to promote cohesion as well as why it is unable to compete for the binding partner. Like *ord[12]*, other *ord* alleles elicit different levels of negative complementation in the two sexes (16). This suggests the possibility that the ORD protein partner (or complex) may be different in males and females.

Another new missense allele, *ord[11]*, also displays interesting negative complementation behavior. The *ord[11]* mutation results in moderate levels of missegregation when placed over a deficiency, and *ord[11]* can be poisoned by stronger missense mutations (Bickel and Orr-Weaver, unpublished observations). In addition, in both males and females, *ord[11]* also shows negative complementation with the

TABLE 5.2. Sex chromosome missegregation in *ord* females.

Genotype[a]	Regular ova (%)	Exceptional ova (%)		Adjusted total[b]	Total % missegregation
	X	O	XX		
ord[12]/ord[10]	44.0	32.9	23.1	1161	56.0
ord[12]/ord[8]	88.8	4.1	7.1	5279	11.2
ord[11]/ord[8]	84.7	7.0	8.3	1478	15.3

[a]Females were crossed to attached-XY, $v\ f\ B$ males.
[b]The progeny total is adjusted to correct for recovery of only half the exceptional progeny.

weaker allele, ord^8 (Tables 5.1 and 5.2). Therefore, ord^{11} represents a special kind of mutation that is capable of poisoning a weak allele but also can be poisoned by a stronger allele. The ord^6 mutation also exhibits this dual character (Bickel and Orr-Weaver, unpublished observations). Further examination of these alleles at the biochemical level may provide critical information about protein-binding versus cohesion activity of the ORD protein.

Consequences of Negative Complementation on Genetic Screens

Interestingly, 10 of the 12 characterized *ord* mutations fall within the C-terminal half of the protein (see Fig. 5.3). This is a much higher frequency than predicted for random hits, especially because it is clear that a mutation in the N-terminus can severely affect ORD function. Genetic screens to isolate new *ord* alleles were carried out by asking for failure to complement ord^1, an allele that exhibits strong negative complementation effects. Therefore, it is possible that such screens were biased toward the isolation of C-terminal lesions, because it appears that negative complementation behavior requires missense mutations in this part of the protein.

A major advantage of screening for failure to complement ord^1 has been the isolation of weak alleles. When placed over a deficiency, alleles such as ord^4 and ord^8 result in low levels of missegregation. These mutations most likely would not have been recovered in a screen that utilized a simple loss-of-function allele on the tester chromosome.

Searching for Proteins That Interact with ORD

One powerful strategy to identify gene products participating in the same cell function is to screen for mutations at additional loci that fail to complement existing known mutations. Such mutations are termed second-site noncomplementers. Second-site noncomplementation has been observed for genes that encode proteins known to physically associate and also those which regulate a common developmental pathway (21, 22). Such noncomplementation may result from structural interactions between aberrant gene products in which a nonfunctional complex is formed. Alternatively, dosage effects may arise because decreasing the amount of one or both gene products inhibits complex formation. A major advantage of screens utilizing second-site noncomplementation is that the mutations recovered often display very interesting phenotypes by themselves. In addition, second-site noncomplementation can be scored in the F_1 generation, unlike recessive suppressors which must be made homozygous before they can be detected.

The screen in which *ord* alleles 7 through 12 were isolated was designed so that second-site noncomplementers on the third chromosome also could be recovered.

We hoped to identify additional gene products essential for sister-chromatid cohesion by looking for mutations that failed to complement *ord¹*. Unfortunately, no such mutations were recovered. However, the isolation of *ord* alleles with different strengths confirms that the design and execution of the screen allowed various levels of missegregation to be detected.

Because the screen was performed before the *ord* gene was cloned, one dilemma in designing the screen was how to select the starting *ord* allele without any molecular knowledge of the nature of *ord* mutations. Given the negative complementation effects observed for *ord¹*, it is clear now that *ord¹* was well suited for a screen to isolate second-site noncomplementers. Such a screen could potentially uncover mutations within the protein partner required for ORD activity. *ord¹/ord⁺* males are phenotypically wild type; *ord¹* is recessive with respect to segregation. We propose that because its active site is altered, ORD¹ can bind the protein partner efficiently but remains inactive. In *ord¹/ord⁺* males, ORD⁺ binds and functions normally and its activity is sufficient to ensure cohesion even in the presence of ORD¹ (see Fig. 5.4B).

Second-site noncomplementation arising from aberrant protein interactions or dosage effects should have been recoverable using the *ord¹* allele. One possibility is that a second-site mutation in the protein partner would not affect protein binding. However, although the complex could form, the mutant partner would not be able to activate ORD's cohesion function (see Fig. 5.4C). This could cause defects in cohesion if the amount of active complex dropped below a threshold level. Alternatively, a second-site mutation that decreased the partner's affinity for ORD might decrease complex formation, resulting in cohesion defects.

Several explanations might account for the failure to obtain second-site noncomplementers of *ord*. Second-site noncomplementation resulting from interactions between two aberrant proteins would be expected to be allele specific and therefore rare. The number of mutants that would need to be screened so as to isolate such a special allele very well might exceed the practical limits of a meiotic missegregation screen. It is also possible that levels of missegregation resulting from second-site noncomplementation were too low to detect in this screen, or alternatively that the combination of the two mutations disrupted meiosis so severely that sterility resulted.

Molecular and biochemical methods offer additional opportunities to identify the protein partner that we propose ORD requires to promote cohesion. Isolation of ORD's binding partner and determination of the requirements for ORD binding will be critical in unraveling the sequence of events that must take place for proper cohesion to be established and modulated during the meiotic divisions.

Acknowledgments. We thank L. Elfring and A. Page for comments on the manuscript. S.B. was a Damon Runyon-Walter Winchell Research Cancer Fund Postdoctoral Fellow. This work was supported by grants from the American Cancer Society and The Council for Tobacco Research, Inc.

References

1. Bickel SE, Orr-Weaver TL. Holding chromatids together to ensure they go their separate ways. Bioessays 1996;18:293–300.
2. Kerrebrock AW, Miyazaki WY, Birnby D, Orr-Weaver TL. The *Drosophila mei-S332* gene promotes sister-chromatid cohesion in meiosis following kinetochore differentiation. Genetics 1992;130:827–41.
3. Goldstein LSB. Mechanisms of chromosome orientation revealed by two meiotic mutants in *Drosophila melanogaster.* Chromosoma (Berl) 1980;78:79–111.
4. Murray AW, Szostak JW. Chromosome segregation in mitosis and meiosis. Annu Rev Cell Biol 1985;1:289–315.
5. Dobson MJ, Pearlman RE, Karaiskakis A, Spyropoulos B, Moens PB. Synaptonemal complex proteins: occurrence, epitope mapping, and chromosome disjunction. J Cell Sci 1994;107:2749–60.
6. Cooke C, Heck M, Earnshaw W. The inner centromere protein (INCENP) antigens: movement from inner centromere to midbody during mitosis. J Cell Biol 1987; 105:2053–67.
7. Earnshaw W, Bernat R. Chromosomal passengers: toward an integrated view of mitosis. Chromosoma (Berl) 1991;100:139–46.
8. Earnshaw WC, Cooke CA. Analysis of the distribution of the INCENPs throughout mitosis reveals the existence of a pathway of structural changes in the chromosomes during metaphase and early events in cleavage furrow formation. J Cell Sci 1991; 98:443–61.
9. Rattner JB, Kingwell BG, Fritzler MJ. Detection of distinct structural domains within the primary constriction using autoantibodies. Chromosoma (Berl) 1988;96:360–7.
10. Theurkauf WE, Hawley RS. Meiotic spindle assembly in *Drosophila* females: behavior of nonexchange chromosomes and the effects of mutations in the *nod* kinesin-like protein. J Cell Biol 1992;116:1167–80.
11. Page AW, Orr-Weaver TL. Activation of the meiotic divisions in *Drosophila* oocytes. Dev Biol 1997;183:195–207.
12. Mason JM. Orientation disruptor (*ord*): a recombination-defective and disjunction-defective meiotic mutant in *Drosophila melanogaster.* Genetics 1976;84:545–72.
13. Davis B. Genetic analysis of a meiotic mutant resulting in precocious sister-centromere separation in *Drosophila melanogaster.* Mol Gen Genet 1971;113:251–72.
14. Miyazaki WY, Orr-Weaver TL. Sister-chromatid misbehavior in *Drosophila ord* mutants. Genetics 1992;132:1047–61.
15. Kerrebrock AW, Moore DP, Wu JS, Orr-Weaver TL. MEI-S332, a *Drosophila* protein required for sister-chromatid cohesion, can localize to meiotic centromere regions. Cell 1995;83:247–56.
16. Bickel SE, Wyman DW, Miyazaki WY, Moore DP, Orr-Weaver TL. Identification of ORD, a Drosophila protein essential for sister-chromatid cohesion. EMBO J 1996; 15:1451–9.
17. Fincham JRS. Genetic complementation. New York: Benjamin, 1966.
18. Foster GG. Negative complementation at the *Notch* locus of *Drosophila melanogaster.* Genetics 1975;81:99–120.
19. Portin P. Allelic negative complementation at the *Abruptex* locus of *Drosophila melanogaster.* Genetics 1975;81:121–33.
20. Raz E, Schejter ED, Shilo BZ. Interallelic complementation among the DER/*flb* al-

leles: implications for the mechanism of signal transduction by receptor-tyrosine kinases. Genetics 1991;129:191–201.

21. Stearns T, Botstein D. Unlinked noncomplementation: isolation of new conditional-lethal mutations in each of the tubulin genes of *Saccharomyces cerevisiae*. Genetics 1988;119:249–60.

22. Xu T, Rebay I, Fleming RJ, Scottgale TN, Artavanis-Tsakonas S. The *Notch* locus and the genetic circuitry involved in early *Drosophila* neurogenesis. Genes Dev 1990; 4:464–75.

6

Role of Cell-Cycle Genes in the Regulation of Mammalian Meiosis

DEBRA J. WOLGEMUTH, VALERIE BESSET, DONG LIU, QI ZHANG, AND KUNSOO RHEE

Mammalian germ cell development is characterized by unique morphogenetic and temporal features that are distinct from those found in somatic cells and which differ between the two sexes. Our working hypothesis is that there are, concomitantly, unique genetic controls of the mitotic and meiotic divisions that occur during germ cell differentiation. To this end, we have used two approaches to elucidate the genes involved in these processes. We have identified the murine homologs and characterized the expression of several of the known regulators of the cell cycle in somatic cells during oogenesis and spermatogenesis, including members of the cyclin, *Cdk* (cyclin-dependent kinase), and *Cdc25* gene families. Some of the regulators functioning in the somatic cell cycle were shown to also be expressed in the germ cell cycle; however, germ cell-specific patterns of expression were observed. Moreover, some cell-cycle-related genes were expressed in cells that were no longer dividing.

We have also initiated efforts to clone previously unidentified genes, some of which may in fact be germ line specific, that may be involved with meiosis. In one such series of experiments, we screened a testis cDNA library under reduced stringency with probes for already characterized members of the *Cdk* family genes. Among the genes that have been identified were the mouse homologue of the *Aspergillus nimA* gene, designated *Nek2*, and a novel *Cdk* family member designated *Pftaire*. The expression and subcellular localization of Nek2 in particular suggested its possible function during meiosis.

In this chapter, we present highlights of some of these observations, in particular noting some of the striking differences we have observed in the expression of several genes between the male and female germ line lineages.

Exploring the Possible Unique Germ Line Function of Known Cell-Cycle Regulators

Cyclins

Since their original discovery in marine invertebrates, cyclins have been identified in a variety of organisms ranging from yeast to man, on the basis of amino acid homology and by functional complementation of yeast cell-cycle mutants (1, 2). Cyclins associate with a catalytic subunit, a cyclin-dependent serine/threonine protein kinase (Cdk). Cyclin B and Cdc2 form a complex known as maturation or mitotic-phase (M-phase) promoting factor, MPF. The kinase activity of MPF induces entry of cells into the M-phase of mitotic as well as meiotic cell cycles. The association of cyclin with Cdc2 initiates a series of alterations in the phosphorylation state of Cdc2, which results in its activation. Destruction of the cyclin protein occurs via the ubiquitin-mediated proteolysis pathway and has been shown to be required for exit of the cell from M-phase (3, 4).

To date, at least nine classes of cyclin genes, designated *cyclin A* to *cyclin I*, have been identified in mammals (2). There are also multiple members of the different classes of cyclins. For example, there may be as many as nine mouse *cyclin B* genes (5, 6), and our own studies have demonstrated at least two mammalian *cyclin A* genes (7, 8). Only a few of the cyclin genes have been studied in any detail in the germ line, and until recently most of these data were limited to oogenesis.

Cyclin A

In vivo data on the functions of *cyclin A* are scarce. During early embryogenesis in *Drosophila*, cells deficient in *cyclin A*, which still express *cyclin B*, cannot enter mitosis (9). In addition, although *cyclin A* mRNA accumulates in *Drosophila* eggs late during oogenesis, the transcripts appear to be synthesized by the nurse cells (10), suggesting a somatic origin. In *Xenopus*, two cyclin As have been identified (11). Maternally produced cyclin A1 and cyclin A2 are both present in oocytes and early embryos, but both are degraded abruptly at the onset of gastrulation. Unlike *cyclin A2*, whose zygotic expression was found throughout embryonic development and in adult tissues as well as in cultured cell lines, *cyclin A1* expression is limited to the germ lines. Although cyclin A is present in *Xenopus* oocytes, its synthesis is not required for meiosis, but may be utilized for early embryonic mitotic events. The role of cyclin A in mitotic versus meiotic events is therefore unresolved.

Cloning of the mouse homologues of the human *cyclin A* gene revealed the presence of at least two members of this gene family (7, 8). In the mouse, *Cyclin A1 (CycA1)* expression is limited to the adult testis, while *Cyclin A2 (CycA2)* is expressed in a wide range of adult tissues, including the testis and ovary. Northern and in situ hybridization analysis showed that *CycA1* expression is limited to spermatocytes late in the pachytene and diplotene stages of meiosis I, while *CycA2* expression was highest in spermatogonia of prepubertal mouse testis and in spermatogonia and preleptotene spermatocytes in adult testis (7, 8). Immunocytochemical localization of CycA1 protein in the adult testis showed CycA1 protein in late diplotene spermatocytes and in metaphase I spermatocytes (Ravnik and Wolgemuth, unpublished observations). In contrast, CycA2 protein was clearly present in spermatogonia and preleptotene, but not in later, meiotic prophase cells (7).

While *CycA1* mRNA could not be detected in oocytes of any stage (pre- or postmeiotic), *CycA2* was readily detected in oocytes in the adult ovary and in ovulated oocytes within the oviductal ampulla (Ravnik and Wolgemuth, unpublished observations). *CycA2* was also shown to be expressed in granulosa cells surrounding the oocyte in growing ovarian follicles. Correspondingly, no CycA1 protein was detected in oocytes in adult ovaries by immunohistochemistry, and only very low levels of CycA1 could be detected in oocytes by immunofluorescence (8). In contrast, readily detectable levels of CycA2 protein were observed in ovarian oocytes and in association with the spindle of the second meiotic division (Ravnik and Wolgemuth, unpublished observations). Thus, the two *Cyclin A* genes are expressed in germ cells in quite different patterns, suggesting that they have distinct functions in the mitotic and meiotic divisions of male and female germ cells. Although *CycA1* appears to be male meiosis specific, *CycA2* clearly is present in both somatic and germinal cells, and its expression patterns further suggest roles in both mitotic and meiotic cell cycles. Furthermore, targeted mutagenesis of the CycA2 gene has recently been shown to result in early embryonic lethality (12).

Cyclin B

As mentioned earlier, there are multiple cyclin B proteins (2), the two best studied of which are vertebrate cyclin B1 (CycB1) and cyclin B2 (CycB2). Biochemical observations highlight the role of cyclin B in regulation of M-phase, but little is known about the in vivo regulation and action of cyclin Bs. Given the essential role of the B cyclins during M-phase and the probability of differential regulation of mitotic and meiotic cell cycles between males and females, we studied the expression patterns of *CycB1* and *CycB2* in the testis and ovary (13–15). Both *CycB1* and *CycB2* were readily detected in the ovary and testis, and both also appeared to be germ cell specific in the testis. In the ovary, both cyclins are present in the somatic granulosa cells as well as oocytes.

Localization of *CycB1* and *CycB2* transcripts by in situ hybridization revealed that, in the testis, *CycB1* is present predominately in the haploid, postmeiotic round spermatids. This result was surprising because these cells have completed M-phase. However, further analysis of CycB1 protein and associated kinase activity showed that CycB1 protein and activity are highest in the pachytene (prophase I) spermatocytes, suggesting that the *CycB1* transcripts present in round spermatids may not be translated efficiently (15). Although these data are consistent with known biochemical data showing CycB1 function at M-phase, it is interesting to note that some kinase activity was detected in elongating spermatids. Because spermatids at this stage are undergoing profound morphogenetic changes, such as replacement of the histones with protamines, perhaps Cdc2 kinase activity is involved in regulatory phosphorylations of proteins involved in spermiogenesis. *CycB2* transcripts were present at highest levels in late pachytene spermatocytes, just before the first meiotic divisions, consistent with a role in M-phase. In the ovary, in situ hybridization showed that while both *CycB1* and *CycB2* are expressed in this organ, the highest level of *CycB1* expression was in growing oocytes, while *CycB2* was expressed uniformly in oocytes and surrounding granulosa cells.

Cyclin D

Three *cyclin D* gene family members have been identified to date: *cyclins D1, D2,* and *D3 (CycD1, CycD2, CycD3)*. All three can be detected in fibroblast cell lines, albeit at different levels (16). Some level of lineage specificity has been observed for the D-type cyclins, in that only *CycD2* and *CycD3* appear to be expressed in T lymphocytes (17). Although these three D-type cyclins are expressed in a somewhat overlapping, redundant fashion in proliferating tissues, their functions as demonstrated by targeted mutagenesis reveal distinct differences during cell proliferation and differentiation. *CycD1*-deficient mice are viable and fertile, but are consistently smaller in size compared to wild-type littermates and show defects in retina and breast development (18, 19). Recently, *CycD2*-deficient mice have been generated and shown to result in female sterility, owing to the inability of ovarian granulosa cells to proliferate normally in response to follicle-stimulating hormone (FSH) (20). *CycD2*-mutant males display hypoplastic testes. Interestingly, some human ovarian and testicular tumors have been shown to contain high levels of *CycD2* mRNA. This suggests that although a cyclin is essential for the normal development of a tissue, it can contribute to neoplastic growth when overexpressed in the same tissue.

Given the association of the D cyclins with growth control and the complex growth requirements for gametogenesis, we analyzed the expression of *CycD1, CycD2,* and *CycD3* in the testis and ovary [(21); Zhang et al., unpublished observations]. All three D-type cyclins could be detected at both mRNA and protein levels in the testis and ovary, but in distinct and surprising patterns. *CycD1* was detected at highest levels in early postnatal testes, and in the adult it was more abundant in testes from germ cell-deficient mutants. In situ hybridization analyses showed that expression of *CycD1* was primarily, if not exclusively, in Sertoli cells. Immunohistochemical localization of CycD1 revealed that it was highly expressed in spermatogonia throughout the testicular development and in Sertoli cells in adult testis. Because Sertoli cells are normally not dividing in the adult animal, this surprising result indicates that CycD1 may be not only involved in the process of cell proliferation but also plays a role in cell differentiation.

In marked contrast, in situ hybridization did not reveal any distinct cellular localization of *CycD2* transcripts in the testis (21). Localization of CycD2 by immunohistochemistry revealed its presence exclusively in actively dividing spermatogonia. In the ovary, *CycD2* expression was readily observable by in situ hybridization and immunohistochemistry in the granulosa cells of growing follicles and in corpea lutea, consistent with other data showing expression of *CycD2* in actively dividing cells (e.g., see ref. 22). It appears that, in the gonads, *CycD2* may be the only D cyclin expressed in patterns clearly associated with cellular proliferation because *CycD3* also exhibited surprising expression patterns.

Few data are available on the expression and activity or regulation of *CycD3*, but like the other D-type cyclins it is most active during the G_1 phase and is expressed in dividing cells. Recent studies have also shown that CycD3 is complexed with Cdk4 and is highly expressed in differentiated myotubes, indicating a role for CycD3 in maintaining cells in the terminally differentiated state (23). We observed *CycD3* expression in the testis to be highest in the nondividing, round spermatids (21). By immunohistochemistry, CycD3-specific staining was detected in the nuclei of mitotically dividing spermatogonia and at even higher levels in the cytoplasm of elongating spermatids (Zhang et al., unpublished observations). Among the somatic cells, anti-CycD3 antibodies strongly localized to the nuclei of Leydig cells, with some weak staining in the cytoplasm of Sertoli cells. Again, we were surprised at this example of another growth-associated cyclin being expressed in cells with no proliferative activity. In general, these observations may indicate functions not specifically related to cell-cycle progression.

Cyclin-Dependent Kinases

The cyclin-dependent kinases (Cdks) are a family of serine/threonine kinases whose enzymatic activities are induced when associated with their regulatory subunits, the cyclins. More than 10 mammalian genes that share functional or structural characteristics with *Cdc2* have been identified (2). Genes that are structurally similar to the Cdks, but whose cyclin partners are not yet known, have been named by virtue of the amino acid homologies to the conserved domain PSTAIRE, including *Kkialre, Pctaire-1, Pctaire-2, Pctaire-3, Pisslre,* and *Pitslre.*

The Cdks play a master role in the decisions that the cell makes for proliferation. A number of regulatory mechanisms are involved in controlling Cdk activity, among the most basic regulation being their interaction with cyclins (24). A single yeast Cdk can interact with various cyclins in different transition points. In contrast, in mammals there are several Cdks that exhibit specific preference for association with specific cyclins and participate only in limited events of the cell cycle [(25; reviewed in (2)]. Mammalian Cdc2 preferentially forms a complex with B-type cyclins and functions at the G_2/M transition. Cdk2 interacts with cyclins A, B, D, and E, forming complexes believed to be important for G_1/S transition. Cdk4 forms complexes with D-type cyclins and functions at G_1 progression. It has been shown that Cdk5 and Cdk6 can also interact with D-type cyclins. Cdk3 was named a Cdk on the basis of its ability to complement yeast *cdc2* mutants (26), even though its cyclin partners have not been determined (27).

Several recent observations have suggested roles for Cdks other than in cell-cycle regulation. For example, Cdk5 is expressed abundantly in terminally differentiated neural cells (28, 29) and may be involved in neuron migration (30). Our data on Cdk5 expression suggest that it may also be linked to apoptotic cell death during development (31). Only a few studies have been carried out to elucidate the biological functions of Cdks in mammalian germ cell development. We have initiated a study to identify the *Cdk* family genes that are involved in spermatogenesis, using a polymerase chain reaction-(PCR-) based cloning method. Many of the resulting clones corresponded to *Cdc2* and *Pctaire-1*, both of which are expressed abundantly in the testis. Somewhat less frequently, *Cdk2, Cdk3, Cdk4, Cdk5, Pctaire-2, Pctaire-3,* and *Pitslre* clones were found. These results revealed that a number of *Cdk* family genes are involved in male germ cell development.

We also have examined the expression of several *Cdk* family genes (*Cdc2, Cdk2, Cdk4, Pctaire-1,* and *Pctaire-3*) in the testis (32). Our results revealed not only that *Cdk* family genes had striking cellular, lineage, and developmental specificity in their expression but also that their expression did not necessarily coincide with known proliferative activity. *Cdc2* and *Cdk2* expression in male germ cells did appear to correlate with mitotic and meiotic activities. Both *Cdc2* and *Cdk2* were expressed most abundantly in spermatocytes. The highest levels of *Cdc2* transcripts were found in late pachytene to diplotene spermatocytes. These differences in Cdk levels within the meiotic phase of the cell cycle are in contrast with the constant levels of *Cdk* expression observed in yeasts (33), *Drosophila* embryos (34), and mammalian tissue culture cells (36). We also observed expression of the *Cdk* family genes in cells with no proliferative activity, suggesting their involvement in cellular functions other than the progression of the cell cycle. For example, all five *Cdk* family genes were shown to be expressed in adult Sertoli cells, cells that are no longer mitotically active, and the highest expression of *Pctaire-1* and *Pctaire-3* was found in postmeiotic spermatids (32).

Cdc25 Protein Phosphatases

The serine/threonine protein kinase activity of Cdc2 is regulated by both its association with cyclin, the regulatory subunit of MPF, and its state of phosphorylation, particularly on Tyr15 and Thr14. The protein phosphatase responsible for dephosphorylating Tyr 15 and Thr14 is the product of the *cdc25* gene, which was first identified in *S. pombe* (24). In *Drosophila*, two *cdc25* homologues have been identified, *string* and *twine* (36, 37). The expression of *string* is detected in mitotically dividing cells and is required for mitosis, while the expression of *twine* is detected only in germ cells undergoing meiosis. The product of *twine* may be required for specific aspects of meiosis, namely, spindle formation (38).

In mammals, three *cdc25* homologs have been identified, designated *Cdc25A, Cdc25B,* and *Cdc25C* [see discussion in (39)]. In the adult mouse, *Cdc25C* is expressed in the highest levels in the testis, in which its expression appears to be germ cell-abundant, if not specific (39). During spermatogenesis, *Cdc25C* is expressed in late pachytene-diplotene spermatocytes, cells that are about to undergo the meiotic divisions. Thus, the pattern of expression of *Cdc25C* is consistent with a proposed function at G_2-M. The *Cdc25C* transcripts nonetheless continue to be present at high levels in round spermatids, and then decrease in elongating spermatids. The localization of *Cdc25A* mRNA to spermatocytes as well as spermatids (Liu and Wolgemuth, unpublished observations) is similar to that of *Cdc25C*. This raises the question as to whether these two genes could perform different functions at the same cell-cycle stage or whether they might be redundant. In contrast, *Cdc25B* is expressed predominantly in the somatic cells of the testis (39), most of which are no longer proliferating, suggesting that *Cdc25B* may also have a function in differentiating or differentiated cells rather than only in cell-cycle control per se.

In the ovary, the cell lineage specificity of the three *Cdc25* family gene expression is very different from that observed in the testis. In contrast to germ cell-abundant expression in the testis, *Cdc25C* transcripts were most abundant in the cumulus granulosa cells surrounding fully grown oocytes in mature follicles. On the other hand, both *Cdc25A* and *Cdc25B* transcripts were detected in granulosa cells and the growing and mature oocytes in medium and large follicles [(39) Liu and Wolgemuth, unpublished observations]. These results suggest that the three *Cdc25* genes in mouse may have lineage- and development-specific functions in both germinal and somatic components of the gonads, some of which may not be directly linked to cell proliferation.

Identifying New Cell-Cycle Regulators Functioning in the Germ Line

To identify additional protein kinase genes that may be responsible for meiosis-specific cell-cycle events, we screened mouse testis cDNA libraries with a reduced stringency using *Cdk* family genes as probes. Among the clones isolated was a novel *Cdk* family gene, *Pftaire-1* (Rhee, et al., unpublished observations). *Pftaire-1* reveals 50% and 45% homology with *Cdk5* and *Cdc2*, respectively, at the amino acid level. *Pftaire-1* was expressed abundantly in brain and testis. Specific expression of *Pftaire-1* at late spermatocytes in the testis suggests its involvement in meiosis.

We also isolated a potential murine homologue of the *Aspergillus nidulans* gene *nimA* (40), designated *Nek2* (Rhee and Wolgemuth, unpublished observations). NIMA is a Ser/Thr protein kinase, but is distinct from Cdc2, structurally as well as biochemically (41). A role for a NIMA-like activity in vertebrate cells has been documented. Ectopically expressed NIMA has been shown to induce germinal vesicle breakdown in *Xenopus* oocytes (42) and premature mitotic events in HeLa cells (42, 43), independent of Cdc2 activity. So far, three mammalian genes, *Nek1*, *Nek2*, and *Nek3*, have revealed structural homology with *nimA*. However, the *Nek* genes have not successfully rescued *nimA* deficiency in *Aspergillus* nor has the misexpression of *Nek* genes in cultured cells resulted in a phenotype (rev. in 41). Thus, the function of the vertebrate *nimA* homologs remains a mystery, despite the clear critical function of this gene in various organisms.

Mouse *Nek2* was shown to be expressed most abundantly in the testis among adult tissues (Rhee and Wolgemuth, unpublished observations). Its expression in the testis was restricted to the germ cells, with highest levels detected in spermatocytes at pachytene and diplotene stages. Immunohistochemical analysis revealed that Nek2 localized to nuclei, exhibiting a nonuniform distribution within the nucleus. Nek2 appeared to be associated with meiotic chromosomes, an association that was better defined by immunolocalization to hypotonically dispersed meiotic chromosomes. This localization was more apparent in regions of dense chromatin, including the sex vesicle, and was also obvious at some of the chromosome ends. The presence of Nek2 protein was not unique to male germ cells, as it was found in meiotic pachytene stage oocytes as well. Furthermore, in an in vitro experimental setting in which meiotic chromosome condensation was induced with okadaic acid, a concomitant induction of Nek2 kinase activity was observed. The expression of *Nek2* in meiotic prophase is consistent with the hypothesis that, in vivo, *Nek2* is involved in the G_2/M phase transition of the cell cycle, in particular, in events of meiosis.

Summary

In summary, our observations to date (Fig. 6.1) have revealed the presence of unique regulators for the mitotic and meiotic cell cycles in the germ line. Furthermore, our data raise the possibility that, during development of the germ line, cell-cycle-related genes may be involved in other functions rather than cell division per se. A major challenge for the future is to identify additional mammalian homologs of meiosis regulating genes from yeast and other organisms. There are several strategies by which such new genes could be identified, including structural similarity or functional complementation. Our progress in this regard is aided by the explosion of information gleaned from studies in genetically more accessible systems, such as yeast and *Drosophila*, in identifying the mammalian counterparts of genes regulating meiosis as well as mitosis.

Acknowledgments. This work was supported in part by grants from the NIH, HD 05077 (DJW), F32 HD07968 (KR), F32 HD08065 (QZ), the NCI CA13696 (DJW), and The Lalor Foundation (VB).

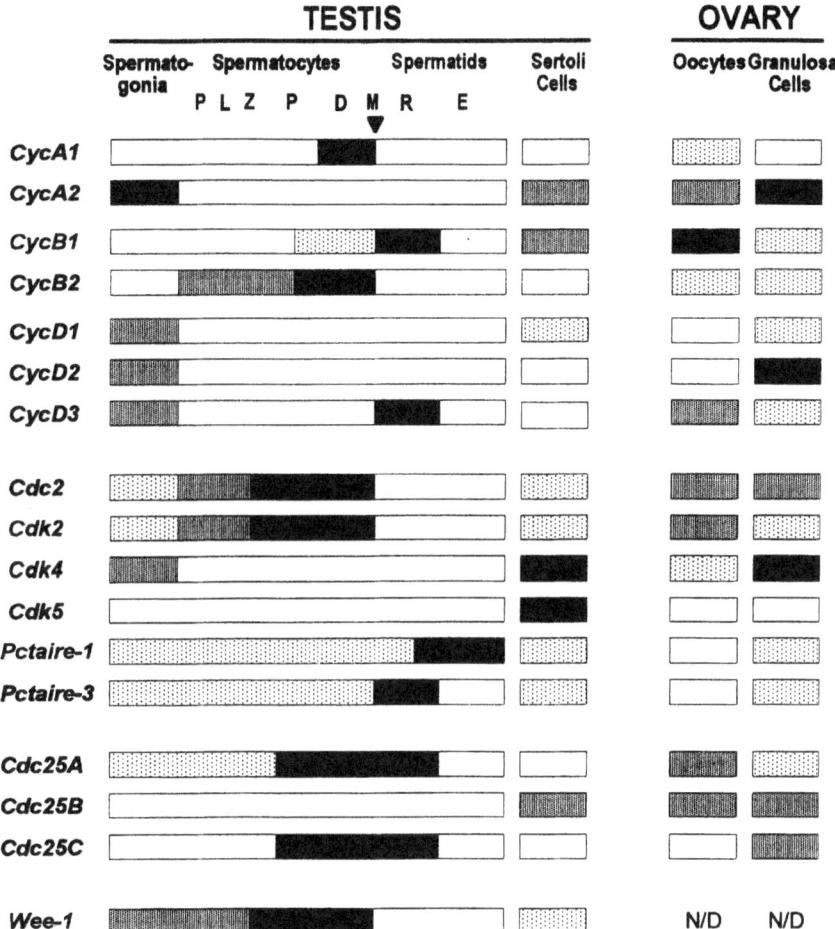

FIGURE 6.1. Summary of the expression of selected cell-cycle regulating genes in the mouse testis and ovary. This figure represents a composite of the results of studies from our laboratory on the levels of expression of the genes denoted on the left. The references containing the data supporting this compilation may be found in the text. The relative levels are depicted by shading of the bars: solid black, most robust expression of a particular gene; white bar, no detection of expression. The data on *Cdk5* expression refer specifically to apparently nonapoptotic cells in the testis and ovary (32). The expression of *CycA1* in the oocyte was only at the protein level and was at extremely low levels [(8); Ravnik and Wolgemuth, unpublished observations].

References

1. Sherr CJ. Mammalian G_1 cyclins. Cell 1993;73:1059–65.
2. Pines J. Cyclins and cyclin-dependent kinases: take your partners. Trends Biochem Sci 1993;18:195–7.
3. Murray AW, Solomon MJ, Kirschner MW. The role of cyclin synthesis and degradation in the control of maturation promoting factor activity. Nature (Lond) 1989;339:280–6.
4. Glotzer M, Murray AW, Kirschner MW. Cyclin is degraded by the ubiquitin pathway. Nature (Lond) 349:132–8.
5. Hanley-Hyde J, Mushinski JF, Sadofsky M, Huppi K, Krall M, Kozak CA, et al. Expression of murine cyclin B1 mRNAs and genetic mapping of related genomic sequences. Genomics 1992;13:1018–30.
6. Lock LF, Pines J, Hunter T, Gilbert DJ, Gopalan G, Jenkins NA, et al. A single cyclin A gene and multiple cyclin B1-related sequences are dispersed in the mouse genome. Genomics 1992;13:415–24.
7. Ravnik SE, Wolgemuth DJ. The developmentally restricted pattern of expression in the male germ line of a murine *Cyclin A, Cyclin A2,* suggests roles in both mitotic and meiotic cell cycles. Dev Biol 1996;173:69–78.
8. Sweeney C, Murphy M, Kubelka M, Ravnik S, Hawkins F, Wolgemuth DJ, et al. A distinct cyclin A is expressed in germ cells in the mouse. Development (Camb) 1996;122:53–64.
9. Lehner CF, O'Farrell PH. The roles of cyclins A and B in mitotic control. Cell 61:535–47.
10. Dalby B, Glover DM. 3'-Non-translated sequences in *Drosophila* cyclin B transcripts direct posterior pole accumulation late in oogenesis and peri-nuclear association in syncytial embryos. Development (Camb) 1992; 115:989–97.
11. Howe JA, Howell M, Hunt T, Newport JW. Identification of a developmental timer regulating the stability of embryonic cyclin A and a new somatic A-type cyclin at gastrulation. Genes Dev 1995;9:1164–76.
12. Murphy M, Stinnakre M-G, Senamaud-Beaufort C, Winston NJ, Sweeney C, Kubelka M, et al. Delayed early embryonic lethality following disruption of the murine cyclin A2 gene. Nat Genet 1997;15:83–6.
13. Chapman DL, Wolgemuth DJ. Identification of a mouse B-type cyclin which exhibits developmentally regulated expression in the germ line. Mol Reprod Dev 1992;33:259–69.
14. Chapman DL, Wolgemuth DJ. Isolation of the murine cyclin B2 cDNA and characterization of the lineage and temporal specificity of expression of the B1 and B2 cyclins during oogenesis, spermatogenesis and early embryogenesis. Development (Camb) 1993;118:229–40.
15. Chapman DL, Wolgemuth DJ. Regulation of M-phase promoting factor activity during development of mouse male germ cells. Dev Biol 1994;165:500–6.
16. Quelle DE, Ashmun RA, Shurtleff SA, Kato JY, Bar-Sagi D, Roussel MF, et al. Overexpression of mouse D-type cyclins accelerates G_1 phase in rodent fibroblasts. Genes Dev 1993;7:1559–71.
17. Ajchenbaum F, Ando K, DeCaprio JA, Griffin JD. Independent regulation of human D-type cyclin gene expression during G_1 phase in primary human T lymphocyctes. J Biol Chem 1993;268:4114–9.
18. Fantl V, Stamp G, Andrews A, Rosewell I, Dickson C. Mice lacking cyclin D1 are

small and show defects in eye and mammary gland development. Genes Dev 1995;9: 2364–72.

19. Sicinski P, Donaher JL, Parker SB, Li T, Fazell A, Gardner H, et al. Cyclin D1 provides a link between development and oncogenesis in the retina and breast. Cell 1995;82: 621–30.

20. Sicinski P, Donaher JA, Geng Y, Parker SB, Gardner H, Park MY, et al. Cyclin D_2 is an FSH-responsive gene involved in gonadal cell proliferation and oncogenesis. Nature (Lond) 1996;384:470–4.

21. Wolgemuth DJ, Rhee K, Ravnik SE. Genetic control of cellular proliferation and differentiation during mammalian gametogenesis. Contracept Fertil Sex 1994;22: 623–6.

22. Matsushime H, Roussel MF, Ashmun RA, Sherr CJ. Colony-stimulating factor 1 regulates novel cyclins during the G_1 phase of the cell cycle. Cell 1991;65:701– 13.

23. Kiess M, Gill, RM, Hamel PA. Expression of the positive regulator of cell cycle progression, cyclin D3, is induced during differentiation of myoblasts into quiescent myotubes. Oncogene 1995;10:159–66.

24. Solomon M J. Activation of the various cyclin/cdc2 protein kinases. Curr Opin Cell Biol 1993;5:180–6.

25. Fang F, Newport JW. Evidence that the G_1-S and G_2-M transitions are controlled by different cdc2 proteins in higher eukaryotes. Cell 1991;66:731–42.

26. Meyerson M, Enders G, Wu C-L, Su L-K, Gorka C, Nelson C, et al. A family of human cdc2-related protein kinases. EMBO J 1992;11:2909–17.

27. van den Heuvel S, Harlow E. Distinct roles for cyclin-dependent kinases in cell cycle control. Science 1993;262:2050–4.

28. Tsai L-H, Takahashi T, Caviness VS, Harlow E. Activity and expression pattern of cyclin-dependent kinase 5 in the embryonic mouse nervous system. Development (Camb) 1993;119:1029–40.

29. Hellmich MR, Pant HC, Wada E, Battey JF. Neuronal cdc2-like kinase: a cdc2-related protein kinase with predominantly neuronal expression. Proc Natl Acad Sci USA 1992;89:10867–71.

30. Nikolic M, Dudeck H, Kwon YT, Ramos YFM, Tsai LH. The cdk5/p35 kinase is essential for neurite outgrowth during neuronal differentiation. Genes Dev 1996;10: 816–25.

31. Zhang Q, Ahuja HS, Zakeri ZF, Wolgemuth DJ. Cyclin-dependent kinase 5 is associated with apoptotic cell death during development and tissue remodeling. Dev Biol 1997:222–33.

32. Rhee K, Wolgemuth DJ. Cdk family genes are expressed not only in dividing but also in terminally differentiated mouse germ cells, suggesting their role possible function during both cell division and differentiation. Dev Dyn 1995;204:406–20.

33. Amon A, Tyers M, Futcher B, Nasmyth K. Mechanisms that help the yeast cell cycle clock tick: G_2 cyclins transcriptionally activate G_2 cyclins and repress G_1 cyclins. Cell 1993;74:993–1007.

34. Edgar BA, Sprenger F, Duronio RJ, Leopold P, O'Farrell PH. Distinct molecular mechanisms regulate cell cycle timing at successive stages of *Drosophila* embryogenesis. Genes Dev 1994;8:440–52.

35. Rosenblatt J, Gu Y, Morgan DO. Human cyclin-dependent kinase 2 is activated during the S and G_2 phases of the cell cycle and associates with cyclin A. Proc Natl Acad Sci USA 1992;89:2824–8.

36. Edgar BA, O'Farrell PH. Genetic control of cell division patterns in the *Drosophila* embryo. Cell 1989;57:177–87.
37. Jimenez J, Alphey L, Nurse P, Glover DM. Complementation of fission yeast *cdc2^{ts}* and *cdc25^{ts}* mutants identifies two cell cycle genes from *Drosophila:* a *cdc2* homologue and *string.* EMBO J 1990;9:3565–71.
38. White-Cooper H, Alphey L, Glover DM. The *cdc25* homologue *twine* is required for only some aspects of the entry into meiosis in *Drosophila.* J Cell Sci 1993;106:1035–44.
39. Wu S, Wolgemuth DJ. The distinct and developmentally regulated patterns of expression of members of the mouse Cdc25 gene family suggest differential functions during gametogenesis. Dev Biol 1995;170:195–206.
40. Osmani SA, Pu RT, Morris NR. Mitotic induction and maintenance by overexpression of a G_2-specific gene that encodes a potential protein kinase. Cell 1988;53:237–44.
41. Fry AM, Nigg EA. The NIMA kinase joins forces with Cdc2. Curr Biol 1995;5:1122–5.
42. Lu KP, Hunter T. Evidence for a NIMA-like mitotic pathway in vertebrate cells. Cell 1995;81:413–24.
43. O'Connell MJ, Norbury C, Nurse P. Premature chromatic condensation upon accumulation of NIMA. EMBO J 1994;13:4926–37.

7

Acquisition of Competence to Enter Meiotic Metaphase During Mammalian Spermatogenesis

Mary Ann Handel and John Cobb

What Is Meiotic Competence?

Little is known about factors that control the exit of the spermatocyte from meiosis I prophase to metaphase I (the G_2/M transition). Before negotiating this critical cell-cycle transition, the spermatocyte must acquire competence to condense and segregate homologous chromosomes and must receive the appropriate signal(s) for entry into metaphase; these two requirements do not necessarily coincide temporally. The work of Wolgemuth and co-workers (see Chapter 6, this volume, for review) has provided ample evidence that a number of important and highly conserved cell-cycle proteins are in the right place at the right time, but it is not known how these proteins relate in either a mechanistic or regulatory sense to acquisition of meiotic competence and the G_2/M cell-cycle transition in spermatocytes.

A significant advantage of using the laboratory mouse for analysis of male meiosis is the ability to isolate, in highly enriched preparations, various cells in the spermatogenic lineage, most notably pachytene spermatocytes (1). These cells, which are in the middle of the first meiotic prophase, are recognized cytologically by the fact that homologous chromosomes are fully paired, with a unique structure, the synaptonemal complex, as a hallmark of that pairing (Fig. 7.1). The dogma of classical genetics tells us that these cells are in the act of recombination, although that view may not be wholly accurate. The earliest stages of meiotic prophase, leptonema and zygonema, when chromosomes first begin to seek and find their homologous pairing partners, last about 1 day each in the mouse. The pachytene stage lasts nearly a week, and is over when chromosomes begin to condense in diplonema and then achieve full condensation of chiasmate bivalents at meiotic metaphase I (MI). At MI, the prior recombination events are visualized cytologically as the chiasmata that link homologues together for proper position-

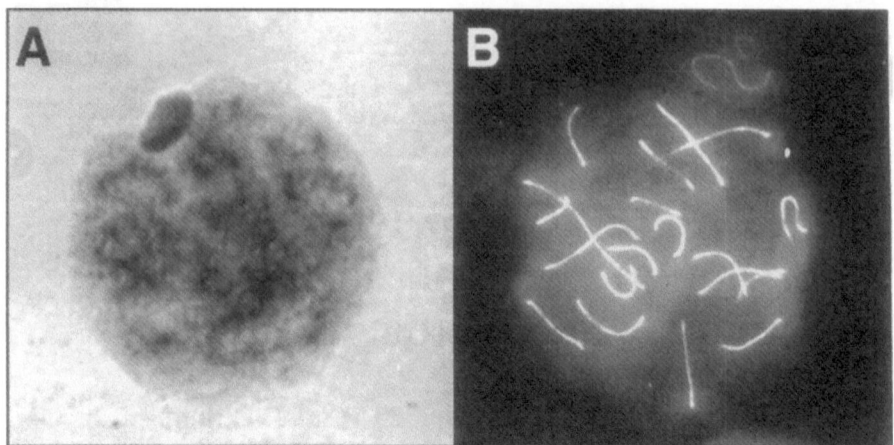

FIGURE 7.1. Typical features of pachytene spermatocytes. (*A*) Air-dried meiotic preparation shows condensed sex body. (*B*) Fixed pachytene spermatocyte nucleus, reacted with a polyclonal antiserum against synaptonemal complex (antiserum kindly provided by Peter Moens), also stained with 4'-6'-diamindno-2-phenylindole (DAPI) to show chromatin.

ing on the meiotic spindle. The importance of these chiasmata for species survival is paramount; they prevent aneuploidy by ensuring accurate segregation of homologous chromosomes at the first meiotic anaphase (2).

Assessing Spermatogenic Meiotic Competence

One barrier to understanding how these important chromosome mechanics relate to regulatory control of the spermatocyte G_2/M transition has been the lack of models for experimental analysis. Recently we turned to short-term cultures of pachytene spermatocytes to study the dynamics of meiotic progress (3). In this approach, spermatocytes are isolated by previously reported methods (1) and cultured for brief periods of time, up to 48 h, in alpha-minimum essential medium (α-MEM) supplemented with 5% fetal bovine serum. Cultured in this manner, pachytene spermatocytes retain the hallmarks of meiotic prophase such as chromosome pairing, presence of SC, and spatial patterns of nuclear transcription, and even make modest progress through the substages of pachynema (3). Because the total length of pachynema exceeds the length of the culture period, we did not expect to see many cells execute the G_2/M transition, and in fact, only 1%–2% of the cells do so (3).

We were curious as to whether pachytene spermatocytes are competent to condense chiasmate bivalents; determining this can inform us about the timing of genetically important meiotic events. To investigate this, we treated cultured cells with okadaic acid (OA), an inhibitor of protein phosphatases previously shown to induce a premature G_2/M transition in oocytes and mitotic cells by activation of the universal cell-cycle protein complex, metaphase-promoting factor (MPF). OA does indeed induce a precocious G_2/M transition in cultured pachytene spermatocytes (4). This was surprising to us because we thought that these cells might have been in the process of recombination and unable to respond to a cell-cycle trigger. To the contrary, cells treated with OA pass through the meiotic stages of diplonema and diakinesis to MI, where chiasmate bivalent chromosomes are visible (Fig. 7.2). Significantly, chiasma count is normal in these cells induced to a premature G_2/M transition (4). Thus, unexpectedly, these results reveal that pachytene spermatocytes are competent to condense metaphase chiasmate bivalents, and they imply that the events of recombination necessary for formation of chiasmate bivalents have either been resolved by pachynema or that they can be rapidly completed given an appropriate cell-cycle stimulus. This in turn suggests that the genetic events of recombination may be tied to regulation of the cell cycle, perhaps by checkpoint surveillance.

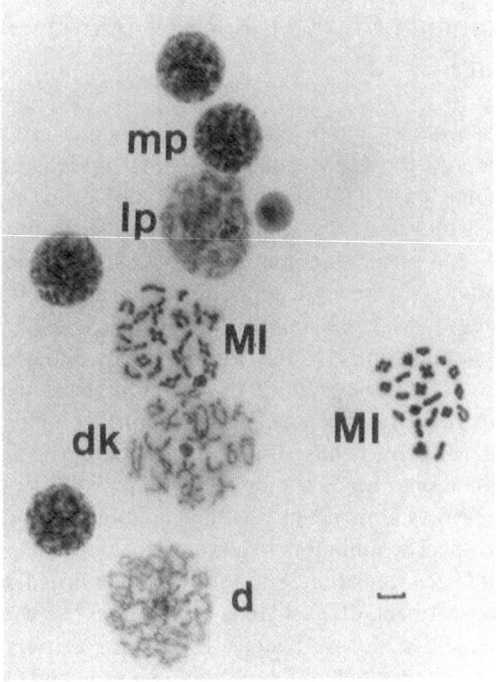

FIGURE 7.2. Bright-field micrograph shows the chromatin condensation induced by okadaic acid (OA) treatment of pachytene spermatocytes. The OA-induced progress through meiotic prophase to metaphase can be visualized in the distinct substages shown here: mp (midpachytene), lp (late pachytene), d (diplotene), dk (diakinesis), and MI (meiotic metaphase I). Bar, 10 μm. Reproduced from Wiltshire et al. (4) with permission of Academic Press.

When during meiotic prophase does competence to condense chiasmate bivalents arise? To answer this question, we cultured spermatocytes at earlier stages of meiotic prophase. We assessed the competence of two populations of cells: one, a mixed population of leptotene and zygotene spermatocytes, and the other, early pachytene spermatocytes, both enriched from testes of juvenile mice. To our surprise, we found that early pachytene spermatocytes are also competent to condense chiasmate bivalent metaphase chromosomes. In contrast, the early prophase leptotene and zygotene spermatocytes responded to OA treatment by condensing chromatin; however, they did not form individualized chromosomes and thus no evidence of chiasmata could be found.

Thus, the spermatocyte acquires chromosomal meiotic competence sometime around the transition from zygonema into pachynema, in the context of the formation of the tripartite SC. Unfortunately, this does not tell us much more about the mechanism of competence acquisition than we might previously have guessed; that is, the ability to resolve recombination event occurs when homologous chromosomes are fully paired.

Protein Requirements for the Induced Meiotic G$_2$/M Transition

What about other aspects of competence? What proteins are required for the spermatocyte to undergo the G$_2$/M transition? One obvious candidate protein complex is MPF, composed of a 34-kDa protein, homologous in all species to the p34 product of the fission yeast cdc2$^+$ gene, and its partner cyclin B. The catalytic component, p34^{cdc2}, is a serine threonine protein kinase whose activity is controlled by its association with the regulatory component, cyclin B, as well as by its phosphorylation status (5–8). MPF activity is typically assessed by measuring its ability to phosphorylate histone H1, and we previously showed OA-stimulated activation of histone H1 kinase activity in lysates of treated pachytene spermatocytes (4). More recently, we have provided further evidence that this is indeed the result of MPF by demonstrating the same activation in the lysate fraction that is adsorbed onto beads conjugated with p13^{suc1}, a protein that binds specifically to the p34^{cdc2} component of MPF (unpublished observations).

We have also used specific inhibitors to learn more about the proteins that play a role in the induced G$_2$/M transition. Staurosporine is a broad-spectrum protein kinase inhibitor. Pretreatment of pachytene spermatocytes with staurosporine blocks the OA-induced G$_2$/M, and histone H1 kinase activity is inhibited in lysates prepared from cells treated with both staurosporine and OA. Pretreatment of cells with olomoucine, which specifically inhibits MPF in vitro (9), did not inhibit the OA-induced transition to metaphase. However, histone H1 kinase activity in cell lysates of olomoucine and OA-treated cells was not inhibited, as it was in the case of staurosporine-treated cells. Taken together, these results imply a role for MPF in the OA-induced cell-cycle transition, but they do not prove this role nor do they exclude the possibility of other proteins as key players. What about the cells that are not competent? We have found that the p34cdc2 protein component of MPF is present in the leptotene and zygotene cells that are not competent to undergo the OA-induced G_2/M transition. Thus the presence of p34cdc2 is not sufficient for the acquisition of competence.

We used inhibitors to demonstrate an absolute requirement for the activity of topoisomerase II in the induced G_2/M transition (10). We found that an inhibitor of topoisomerase I, camptothecin, did not impede the OA-induced progress of pachytene spermatocytes to MI. However, two differently acting inhibitors of topoisomerase II, teniposide and ICRF-193, caused virtually 100% inhibition of the induced G_2/M transition (Fig. 7.3). The mechanism of topoisomerase II enzymatic inhibition by teniposide involves double-strand DNA breaks that are not ligated in the presence of the inhibitor. In contrast, the mechanism of ICRF-193 inhibition is different, and enzyme activity is prevented by a mechanism not leaving DNA strand breaks. Thus, we infer that these results provide evidence for a requirement for topoisomerase II decatenation activity in the condensation of chiasmate chromosomes at the G_2/M transition and not simply a cell-cycle inhibition caused by the presence of DNA strand breaks. Interestingly, the disassembly of the SC that is induced by OA also occurs in the presence of teniposide or ICRF-193; thus, the activity of topoisomerase II is not required for this aspect of

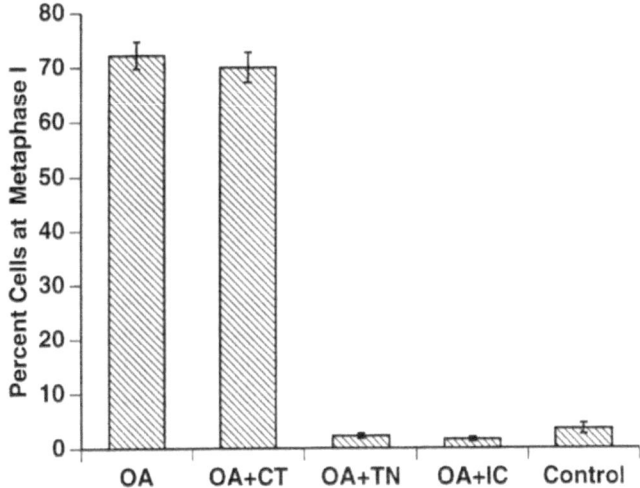

FIGURE 7.3. Effects of topoisomerase inhibitors on the okadaic acid-induced transition of cultured spermatocytes from pachytene to meiotic metaphase I (MI). Aliquots of cell suspensions of pachytene spermatocytes were cultured overnight and then treated for 6 h with 5 μM okadaic acid (OA), alone or in combination with topoisomerase inhibitors. CT, 50 μM camptothecin (topoisomerase I inhibitor); TN, 50 μM teniposide (topoisomerase II inhibitor); IC, 10 μM ICRF-193 (topoisomerase II inhibitor); control, solvent alone. Air-dried preparations of cells were scored for numbers of cells at MI. The mean with SEM of repeat experiments is shown.

the meiotic G_2/M transition. Nonetheless, these results demonstrate that topoisomerase II protein and its activity must constitute one aspect of the spermatocyte's competence to undergo the OA-induced G_2/M transition. We found that topoisomerase II transcript and protein are present both in the incompetent leptotene and zygotene spermatocytes and in the competent pachytene spermatocytes (10). Moreover, topoisomerase enzymatic activity, detected by an in vitro kinetoplast decatenation assay, can be detected in both incompetent and competent cells. We have found that topoisomerase II in pachytene spermatocytes is localized to synapsed chromosomes, as might be expected for a protein that is critically involved in chromatin condensation to form individualized chromosomes. It will be interesting to determine if its localization changes during the G_2/M transition; however, preliminary results indicate no dramatic changes. Taken together, these data demonstrate that topoisomerase II activity is necessary but not sufficient for spermatocyte acquisition of meiotic competence.

Summary

The somewhat fortuitous observation that OA induces a precocious G_2/M cell-cycle transition in cultured mouse pachytene spermatocytes has provided an avenue to the analysis of the acquisition of meiotic competence to condense metaphase bivalent chromosomes. We have found that early prophase spermatocytes are not competent to condense chiasmate bivalent chromosomes, whereas pachytene spermatocytes are. Thus, chromosomal meiotic competence arises in the context of the SC and homologous chromosome pairing and probably involves resolution or partial resolution of the molecular events of recombination. We can expect, as we learn more about these events and their temporal pattern of execution in mammalian cells, that we will learn more about how this aspect of competence is acquired. For instance, we might imagine the existence of a meiotic checkpoint that monitors DNA strand breaks. These aspects of chromosomal competence are important, for mature chiasmata are essential to ensure accurate chromosome segregation during the meiotic division phase and thus prevent aneuploid offspring. Meiosis is an event unique to gametogenesis, and these events are required for normal meiosis and gametogenic success; thus they represent attractive targets for contraceptive interference with spermatogenesis.

Spermatocyte acquisition of competence also involves localization and activation of those proteins necessary to enzymatically mediate the events of the meiotic cell-cycle transition, such as condensation of individualized chromosomes and disassembly of the SC. We have demonstrated a requirement for the decatenation activity of topoisomerase II in the condensation of individualized chiasmate bivalent chromosomes at meiotic metaphase, but not in the disassembly of the SC that occurs at the same time. Thus it appears that there are a number of independent meiotic processes that are set in motion at the time of the cell-cycle transition. Furthermore, our studies of the synthesis and activity of topoisomerase II demonstrate that its presence and activity are necessary but not sufficient for the meiotic cell-cycle transition. This is probably a general principle that will be extended to other cell-cycle proteins as we come to learn more about them and their roles in meiosis.

The model that we have developed for analysis of competence by premature, pharmacological induction of the meiotic G_2/M transition is seductive, and it is all too tempting to extrapolate conclusions to the transition from pachynema to diplonema and metaphase I that occurs in the seminiferous tubule. Thus, it is essential to remember that, although we can demonstrate acquisition of a degree of meiotic competence by pachytene spermatocytes by this method, the cells we study in culture would not normally undergo the G_2/M transition for a number of days. We have essentially no information about the endogenous testicular regulators of the meiotic cell cycle, or the checkpoint surveillance mechanisms that are operable, or the role of the "nonmeiotic" but "spermatogenic" growth and differentiation that occurs during meiotic prophase. Clearly, our future work is cut out for us, and it will be a busy one!

Acknowledgments. This work was supported by grants from the NIH (HD31376 and HD33816) to M.A.H., and by a predoctoral fellowship from the NSF to J.C. We are grateful to Cynthia Park for technical assistance and to Drs. John Eppig, Bruce McKee, and Laura Richardson for comments on the manuscript.

References

1. Bellvé AR. Purification, culture and fractionation of spermatogenic cells. Methods Enzymol 1993;225:84–113.
2. Koehler KE, Hawley RS. Sherman S, Hassold T. Recombination and nondisjunction in humans and flies. Hum Mol Genet 1996;5:1495–504.
3. Handel MA, Caldwell KA, Wiltshire T. Culture of pachytene spermatocytes for analysis of meiosis. Dev Genet 1995;16:128–39.
4. Wiltshire T, Park C, Caldwell KA, Handel MA. Induced premature G_2/M transition in pachytene spermatocytes includes events unique to meiosis. Dev Biol 1995;169:557–67.
5. Lewin B. Driving the cell cycle: M phase kinase, its partners, and substrates. Cell 1990;61:743–52.
6. Murray AW, Hunt T. The cell cycle: an introduction. New York: Freeman and Oxford University Press, 1993.
7. Murray AW, Kirschner MW. Dominoes and clocks: the union of two views of the cell cycle. Science 1989;246:614–21.
8. Nurse P. Universal control mechanism regulating onset of M-phase. Nature (Lond) 1990;344:503–8.
9. Vesely J, Havicek L, Strnad M, Blow JJ, Donella-Deana A, Pinna L, et al. Inhibition of cyclin-dependent kinases by purine analogs. Eur J Biochem 1994;224:771–86.
10. Cobb J, Reddy RK, Park C, Handel MA. Analysis of expression and function of topoisomerase I and II during meiosis in male mice. Mol Reprod Dev 1997;46:489–98.

8

Regulated Synthesis and Role of DNA Methyltransferase During Meiosis

JACQUETTA M. TRASLER, CARMEN MERTINEIT, AND
TONIA E. DOERKSEN

In mammals, the methylation of cytosine residues in DNA is postulated to be involved in a number of processes including gene regulation, development, X-chromosome inactivation, genomic imprinting, and carcinogenesis. Sex- and sequence-specific patterns of DNA methylation are established in the germ line (1–3) and further modified during embryogenesis. The chemical modification of genes by DNA methylation provides a way in which genes can be turned on or off at specific times. Overall, the sperm genome is more methylated than that of the oocyte (4). Methylation of DNA is one of the major candidates proposed to mark the mother's and father's genes differently, in the process of genomic imprinting, which is also initiated in the germ line (5, 6). Our objectives are to determine the mechanisms by which DNA methylation patterns are established during spermatogenesis and the impact on the early embryo of disrupting DNA methylation in male germ cells. Our results to date indicate that DNA methylation is highly regulated in the germ line (1, 7–10) and suggest that decreases in DNA methylation in male germ cells result in alterations in sperm production and abnormalities in early embryo development (11).

DNA Methylation: Advantages and Disadvantages for Mammals

DNA methylation occurs at the 5-position of cytosine at approximately 3×10^7 sites in the mammalian genome, for the most part within CpG dinucleotides (12). It is a postreplication process catalyzed by DNA (cytosine-5)-methyltransferase (DNA MTase). Methylation of CpG sites within promoter sequences of genes almost invariably silences transcription (13) and provides a means of compartmentalizing large genomes into expressed and unexpressed sequences. Once set down (de novo methylation), methylation patterns are clonally inherited (maintenance methylation), allowing gene expression information to be transmitted at the time of DNA replication and cell division (14).

Disruption of the one known form of DNA MTase by gene targeting has underscored the essential nature of DNA methylation in mammals. Homozygous mutants die at midgestation (15, 16) and show evidence of developmental asynchrony (17), indicating that DNA methylation is essential for mammals. The Xist gene, normally expressed only from the inactive X chromosome, is expressed in the mutants from both Xs in females and the one X in males, demonstrating a role for DNA methylation in X inactivation (18). DNA methylation fulfills the four requirements of the biochemical modification of DNA or chromatin that account for imprinting. The modification must (i) be made before fertilization, (ii) be able to confer transcriptional silencing, (iii) be stably transmitted through mitosis in somatic cells, and (iv) be reversible on passage through the opposite parental germline (6). Studies in the DNA MTase-deficient mice provided strong support for the theory that allele-specific methylation underlies the process of genomic imprinting because three imprinted genes (Igf2 receptor, Igf2, and H19), which are normally expressed from only one allele, were either expressed from both alleles or neither allele in homozygous mice (19). Using a 'knock-in' procedure to insert DNA MTase into one of the mutant alleles, Tucker et al. (20) demonstrated that methylation at imprinted alleles was not complete unless the genes had been transmitted through the germ line.

DNA methylation, although important for gene expression and development, is also dangerous (21). Sites of methylation in DNA are mutagenic 'hot spots' because 5-methylcytosine spontaneously deaminates to thymine, giving rise to $C \rightarrow T$ transition mutations. This phenomenon is reflected in the high frequency of mutations at CpG sites in DNA, accounting for about one-third of all mutations in humans (21, 22). Random errors in methylation or 'ectopic' methylation are also a problem. For example, the association of DNA methylation errors with cancer is well documented (23), including germ line mutations in tumor suppressor genes (increased methylation, decreased gene expression), alterations in overall DNA methylation patterns, and high sustained levels of DNA MTase activity.

These examples underscore the importance of DNA methylation and the regulation of DNA MTase activity. Little is known, however, about the factors or mechanisms regulating DNA MTase or the mechanisms by which altered levels of DNA MTase or DNA methylation affect development. An important first step is to address the role of DNA methylation in gametogenesis and the programming of the genome for expression in the embryo.

Creating and Maintaining DNA Methylation Patterns in the Germ Line

How are DNA methylation patterns initially set up in the genome? DNA is thought to be mostly unmethylated at two developmental stages, the primordial germ cell and the blastocyst (2, 4). Current data support two main 'waves' of de novo genomic methylation that establish new methylation patterns during gametogenesis and early postnatal development. Once established, patterns can be maintained through clonal inheritance or lost through either a postulated active demethylation process or when methylation does not occur following replication. From the gene knockout studies, DNA MTase is clearly important and is to date the only isolated component of the DNA methylation system. Only one DNA MTase enzyme, a monomeric protein with an apparent M_r of 190,000, encoded by the *Dnmt* gene, has been found in mammals (24). The mouse DNA MTase cDNA was cloned in 1988 (24) and the human cDNA in 1992 (25); the mouse and human genomic sequences and intron/exon structures have not been published. DNA MTase is capable of methylating both unmethylated DNA (de novo methylation) and hemimethylated DNA (maintenance methylation). Although the intact enzyme prefers hemimethylated DNA as its substrate, cleavage of the N-terminal sequence increases specificity toward unmethylated DNA (26). Somatic cells with differing levels and patterns of genomic methylation contain identical or very similar forms of DNA MTase protein or mRNA. These data support the hypothesis that DNA methylation patterns are modulated by a change in DNA MTase enzyme activity or access to DNA (27).

Our initial studies, on the mouse transition protein 1 and protamine 1 and 2 genes, demonstrated that DNA methylation is a dynamic process in the testis and provided evidence that methylation and demethylation events could occur after replication in male germ cells (1). To determine mechanisms underlying the establishment and maintenance of DNA methylation patterns in the male genome, we have examined the expression, localization, and regulation of DNA MTase during spermatogenesis. Interestingly, DNA MTase mRNA(s) are expressed at higher levels in the testis and ovary than all other tissues in adult mice (10). Immunoblot and enzyme activity assays indicated that DNA MTase is developmentally regulated at transcriptional and posttranscriptional levels in the developing testis (7).

Northern blot analysis of isolated populations of mouse germ cells showed that the 5.2-kb DNA MTase mRNA found in all proliferating cell types is present in type A and B spermatogonia, preleptotene and leptotene/zygotene spermatocytes, and round spermatids (9). In contrast, isolated pachytene spermatocytes contain a novel 6.2-kb DNA MTase transcript (7, 9, 10). Our localization of DNA MTase mRNA to spermatogonia and early spermatocytes is supported by two other studies. Using a reverse transcriptase-polymerase chain reaction assay, Singer-Sam and colleagues (28) detected the presence of high levels of DNA MTase mRNA in spermatogonia and early meiotic (leptotene/zygotene) spermatocytes; the study did not analyze transcript size. Numata et al. (29) used in situ hybridization with an oligonucleotide probe that detects both the 5.2- and 6.2-kb transcripts and confirmed the presence of high levels of DNA MTase mRNA expression in preleptotene, leptotene, and zygotene spermatocytes. We performed polysome analysis to monitor the translational status of the two DNA MTase transcripts (10). We found that 100% of the 5.2-kb mRNA but only 15% of the 6.2-kb mRNA was associated with polysomes, suggesting that the 6.2-kb transcript is translationally regulated or perhaps less efficiently translated than the 5.2-kb transcript.

Western analysis detected a protein in spermatogenic cells with a relative mass of 180,000–200,000, which is close to the known size of the somatic form of mammalian DNA MTase (7, 9). DNA MTase protein is expressed in isolated mitotic, meiotic, and postmeiotic male germ cells, with the exception of pachytene spermatocytes. At pachytene, the protein is undetectable by immunoblot analysis (7, 9). Furthermore, by immunocytochemistry, DNA MTase is localized to the nuclei of all male germ cells except pachytene spermatocytes (9). A subset of spermatogonia (mitotic) and postreplicative leptotene/zygotene (meiotic prophase) germ cells display prominent nuclear foci that are strongly enriched in DNA MTase (9) (Fig. 8.1). The immunolocalization of DNA MTase in isolated germ cell fractions was confirmed by examining DNA MTase expression on tissue sections of mouse testis at different stages of spermatogenesis (Fig. 8.2). The finding of DNA MTase in postmitotic germ cells was surprising. In cultured somatic cells, the enzyme is expressed predominantly in dividing cells and is present only in nuclei, where it is targeted to replication foci during the S-phase (30). We hypothesize that DNA MTase serves a germ cell-specific function(s) in early meiotic cells that is distinctly different from its role in mitotic cells. We predict that de novo methylation may take place in leptotene/zygotene spermatocytes in the nuclear foci we have observed. De novo methylation in these

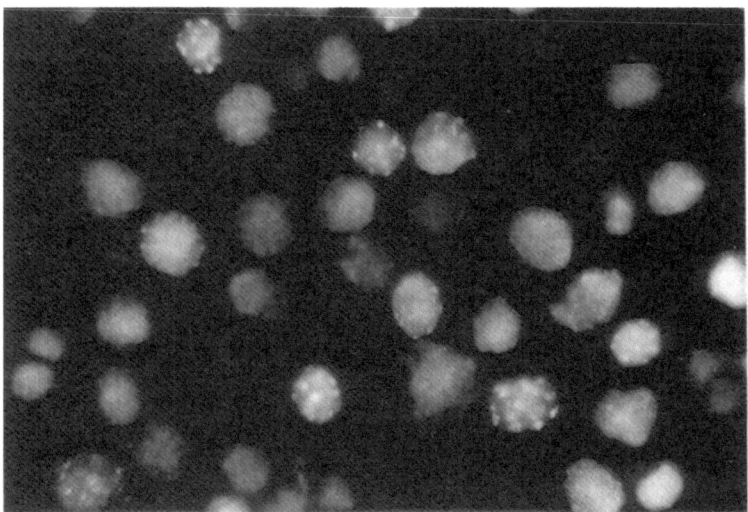

FIGURE 8.1. DNA MTase expression in isolated leptotene/zygotene spermatocytes fixed on slides and incubated with the anti-pATH52 rabbit polyclonal antibody against DNA MTase (9, 26, 30). Immunofluorescence reveals uniform nuclear staining as well as the presence of punctate nuclear foci in some cells.

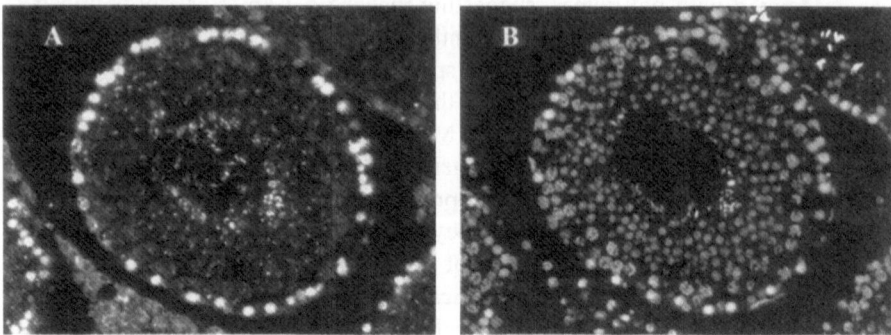

FIGURE 8.2. DNA MTase expression in the mouse testis. (*A*) DNA MTase immunofluorescence (anti-pATH52 antibody) in a seminiferous tubule of the testis at stage VII–VIII of the cycle. Note presence of staining of spermatogonia and preleptotene spermatocytes and absence of staining of pachytene spermatocytes. Examination of staining in tubules at other stages of the cycle confirmed that DNA MTase was expressed in spermatogonia and preleptotene, leptotene, and zygotene spermatocytes but was undetectable in pachytene spermatocytes. (*B*) The same tubule as in (*A*), stained with the nuclear dye propidium iodide to show the position of all cells in the tubule.

cells could be involved in genomic imprinting, male germ line X inactivation, or post-DNA repair methylation.

Parallels in DNA MTase expression between the male and female germ lines may provide clues to the timing and regulation of DNA methylation events. Furthermore, differences in DNA MTase expression patterns might underlie sex-specific methylation of imprinted genes. We are currently using assays of DNA methylation levels with DNA MTase immunocytochemistry to examine the role and localize and compare expression of the enzyme during prenatal and postnatal male and female germ line development. Interestingly, previous studies indicate that the oocyte expresses a novel smaller DNA MTase isoform, not found in male germ cells (31). The molecular basis, role, and regulation of this isoform are unknown.

As mentioned earlier, we showed that DNA MTase is sharply downregulated at pachytene (9). DNA MTase protein was also undetectable in pachytene spermatocyte extracts from Sprague-Dawley rats (8). We suggest that DNA MTase is downregulated at the time of genetic recombination at the pachytene phase of meiosis to protect meiotic chromosomes from ectopic methylation. Our hypothesis is based on the fact that during crossing-over (genetic recombination), several types of DNA structures (four-way junctions, mismatches, and hemimethylated sites) are present (9, 12). These DNA structures are known to be excellent substrates for DNA MTase and could make CpG sites that are not normally methylated vulnerable to methylation (9, 12). Ectopic methylation would be expected to interfere with gene expression during subsequent germ cell development or increase mutation levels in male germ cells. We are currently testing this hypothesis by developing transgenic mice that overexpress DNA MTase in pachytene sper-matocytes. The recent isolation of a new DNA MTase 5'-sequence encoding 171 new amino acids makes such experiments possible (32, 33). The inability of a number of labs to stably overexpress DNA MTase in cell lines, along with our data of a 5'-extended testis-specific 6.2-kb mRNA (9, 10), suggested that the cloned mouse and human DNA MTases might be incomplete. A new 5' sequence for mouse and human DNA MTase, including three upstream exons, was recently reported (32, 33). The new first exon, containing a new ATG, is more than 10 kb upstream of the previously described first exon. Importantly, a minigene contain-ing the new upstream sequence, when expressed, was able to restore methylation in DNA MTase-negative embryonic stem (ES) cells (32). In contrast, constructs containing the original ATG were inefficiently expressed in ES cells.

Manipulating DNA Methylation in the Germ Line

If the DNA methylation patterns taking place in the testis are important, what happens when DNA methylation in this tissue is altered? To address whether decreases in DNA methylation affect male germ cell development and function, we have used 5-azacytidine (11). 5-Azacytidine, a cytidine analog that contains a nitrogen moiety at the 5-position of the pyrimidine ring, where a methyl group is normally added by DNA MTase, is a useful tool for studying DNA methylation. The drug becomes incorporated into DNA during DNA replication and inhibits DNA methylation (34) and DNA MTase activity (35) and may also directly inhibit DNA MTase (36). Treatment with 5-azacytidine results in active transcription of previously silent cellular genes (37–39) and alters the differentiated state of cultured cells (40). Administration of 5-azacytidine to rodents for 1 year caused testicular tumors (41, 42) but the nondemethylating analog 6-azacytidine did not. 6-Azacytidine has a nitrogen in the 6-position of the pyrimidine ring, leaving the 5-position open for addition of a methyl group, and has been used in a number of studies to control for cytotoxic effects of 5-azacytidine.

A low-dose chronic treatment regimen was used in which adult male Sprague-Dawley rats ($n = 4–8$ per group) were administered 0 (saline), 2.5 (low dose), and 4.0 or 5.0 (high dose) mg/kg of 5-azacytidine or the control drug 6-azacytidine by intraperitoneal (i.p.) injection, three times a week for 4 and 11 weeks (11). The length of the two treatments was designed to expose male germ cells at different times during their development (Fig. 8.3). The kinetics of spermatogenesis in the rat assured that 4 weeks of drug treatment exposed germ cells during spermio-genesis and epididymal transit, whereas 11 weeks of treatment exposed sper-matocytes and spermatogonia in addition to more mature germ cells (43). At the end of the treatment, males were mated to determine effects on fertility and embryo development and effects on the male reproductive system were assessed. To analyze effects of male treatment on the offspring, each male was mated overnight with two female rats in proestrous. Success of mating was determined the next morning by examination of vaginal smears for the presence of sper-

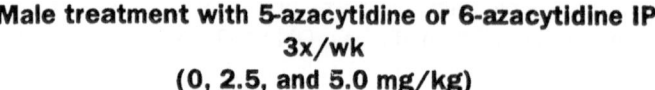

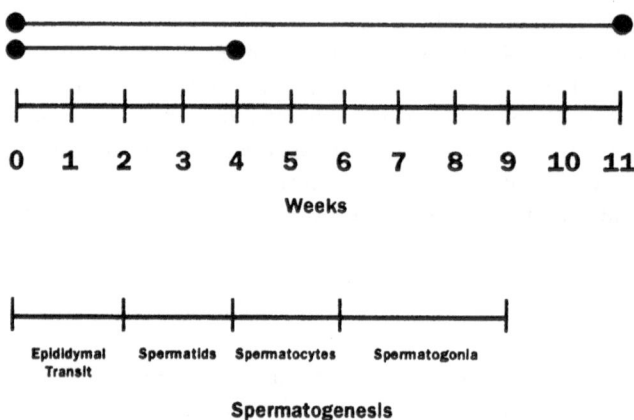

FIGURE 8.3. Experimental design of the 5-azacytidine and 6-azacytidine experiments. Male rats were treated for 4 or 11 weeks, as indicated by lines between large dots (large dots indicate the beginning and end of each treatment). The lower part of the diagram indicates the germ cell types affected by the 4- and 11-week treatments.

matozoa. Females were killed on Day 20 of gestation and cesarian sections were performed. The ovaries were removed and the number of corpora lutea, representing the number of eggs ovulated, were counted. The numbers of implantations, resorbed fetuses, and live fetuses were examined. Pregnancy rate (number of sperm-positive females that became pregnant), preimplantation loss (difference between the number of eggs ovulated and the number of implantation sites), postimplantation loss (number of resorbed or dead fetuses), and incidence of small or malformed fetuses were determined. To examine earlier embryos, males treated with 5-azacytidine for 16 weeks were mated with females. The females were killed on day 2 of gestation. Embryos were flushed from the oviducts and assessed for abnormalities.

Rats in all groups gained weight throughout the treatment, appeared healthy, and demonstrated normal mating ability, suggesting minimal effects of the drugs on the general health of the animals. 6-Azacytidine did not affect the males or their offspring (11). In contrast, with 5-azacytidine, there were marked time-dependent, dose-related decreases in fertility and increases in early embryo loss (Tables 8.1 and 8.2). As indicated by alterations in testis weight and germ cell

TABLE 8.1. Effect of 5-azacytidine treatment on the male reproductive system.

Treatment group	Testis weight[a]		Testicular sperm counts[b]	
	4 Weeks	11 Weeks	4 Weeks	11 Weeks
Saline	3.4 ± .1	3.7 ± .1	186 ± 4	198 ± 5
2.5 mg/kg	3.4 ± .1	3.0 ± .1*	188 ± 6	163 ± 13
4.0/5.0 mg/kg	3.1 ± .1	2.1 ± .1*	176 ± 10	42 ± 13*

[a]Values represent mean in grams ± SEM.
[b]Values represent mean number of sperm × 10^{-6} ± SEM.
*Denotes significance $p \leq .05$ vs. control (saline) values.

numbers or death of embryos, postmeiotic cells (4-week treatment) were un-affected. Treatment of males with 5-azacytidine for 11 weeks resulted in dose-dependent reductions in testis and epididymal weights and sperm counts (Table 8.1). In addition, both doses of 5-azacytidine resulted in significant increases in preimplantation loss with the high dose also causing a decrease in fertility (Table 8.2). No increase in postimplantation loss or fetal malformation was seen. When early (2-day-old) preimplantation embryos sired by males treated with 5-azacytidine or saline for 16 weeks were examined, the total numbers of embryos flushed were similar in the control and treated groups, as was the average number of unfertilized oocytes (11). The majority of the embryos in the control and low-dose 5-azacytidine groups were at the 2-cell stage (averaging 6–8 embryos/litter). In contrast, in the high-dose 5-azacytidine group on average only 1 embryo/litter was a normal 2-cell embryo, and there was a striking increase in the number of abnormal embryos (Table 8.2; Fig. 8.4). Morphologically, abnormal embryos were markedly different from normal embryos (see Fig. 8.4). The results suggest that sperm from males treated with 5-azacytidine were functional in fertilizing the oocytes, but not in embryo development, resulting in death of embryos before implantation.

TABLE 8.2. Effect of paternal 5-azacytidine treatment on progeny outcome.

Treatment group	Pregnancy rate (20 Days)[a]		Preimplantation loss (20 Days)[b]		Abnormal early embryos (2 Days)[c]
	4 Weeks	11 Weeks	4 Weeks	11 Weeks	16 Weeks
Saline	100%	88%	13 ± 5%	6 ± 6%	0.3 ± 0.2
2.5 mg/kg	78%	75%	18 ± 6%	29 ± 9%*	2.6 ± 1
4.0/5.0 mg/kg	90%	11%*	15 ± 5%	50% (one litter)	10.1 ± 1*

[a]Percent of sperm-positive females that became pregnant.
[b]Values represent the difference between number of eggs ovulated and the number of implantation sites in percent (mean/litter).
[c]Values represent the average number of abnormal embryos per litter ± SEM.
*Denotes significance $p \leq .05$.

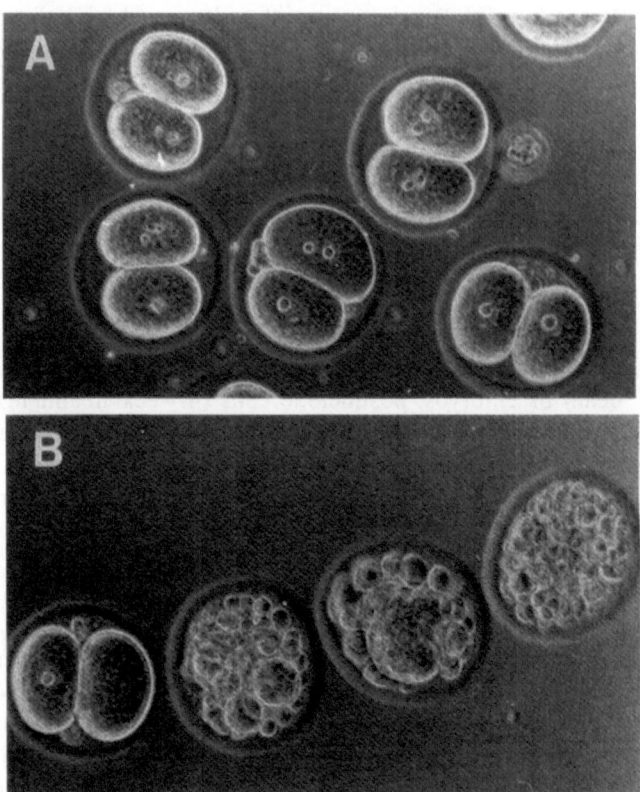

FIGURE 8.4. Light micrographs of day 2 embryos sired by control and 5-azacytidine-treated males. Most of the embryos sired by saline-treated males (*A*) were normal 2-cell embryos, but the majority of the embryos sired by males treated with the high dose of 5-azacytidine (*B*) were markedly abnormal.

Summary

In summary, paternal administration of 5-azacytidine interfered with spermato-genesis and resulted in alterations in fertilization and early embryo development. Lack of an effect of 5-azacytidine after 4 weeks of treatment suggests that there are minimal effects on spermatids and spermatozoa. Effects after 11–16 weeks of treatment could result from the incorporation of 5-azacytidine into replicating DNA during mitosis in spermatogonia or during repair synthesis in spermatocytes or both. Alternatively, the drug may inhibit DNA MTase in spermatogonia or spermatocytes; both cell types exhibit high levels of the enzyme (9, 28, 29). As one mechanism for the male germ cell and embryo abnormalities, we postulate that 5-azacytidine may induce alterations in germ cell DNA methylation patterns. In support of this, 6-azacytidine, which is not known to alter DNA methylation, did not affect the male reproductive system or the progeny outcome. Furthermore, 16 weeks of treatment with the high dose of 5-azacytidine is associated with a 30% decrease in overall DNA methylation in sperm (Doerksen and Trasler, unpublished observations). Together our studies provide evidence of a link between DNA methylation and altered fertility and abnormal embryo development and point to the germ cell types involved. We are currently pursuing the mechanisms underlying these effects. Based on our results to date, we suggest that the drug interferes with the establishment or maintenance of DNA methylation patterns in male germ cells and subsequently alters the normal expression of spermatogenesis-specific genes or other processes such as germline X inactivation, DNA repair, or genomic imprinting.

Acknowledgments. This work was supported by a grant from the Medical Research Council of Canada and an FCAR Team Grant. J.M.T. is a Scholar of the Fonds de la Recherche en Santé du Québec (FRSQ). T.E.D. and C.M. are supported by graduate student scholarships from the Montreal Children's Hospital Research Institute and the McGill Faculty of Medicine. The authors thank Dr. T. Bestor for the mouse DNA MTase cDNA probe, the DNA MTase antibody (anti-pATH52), and many helpful discussions; Dr. D. Laird for help with the confocal microscopy; Ms. G. Benoit and Mr. E. Simard for technical assistance; and Mr. L. Ehler and Mr. A. Forster for photography and artwork.

References

1. Trasler JM, Hake LE, Johnson PA, Alcivar AA, Millette CF, Hecht NB. DNA methylation and demethylation events during meiotic prophase in the mouse testis. Mol Cell Biol 1990;10:1828–34.
2. Chaillet JR, Vogt TF, Beier DR, Leder P. Parental-specific methylation of an imprinted transgene is established during gametogenesis and progressively changes during embryogenesis. Cell 1991;66:77–83.
3. Kafri T, Ariel M, Brandeis M, Shemer R, Urven L, McCarrey J, et al. Developmental pattern of gene-specific DNA methylation in the mouse embryo and germ line. Genes Dev 1992;6:705–14.
4. Monk M, Boubelik M, Lehnert S. Temporal and regional changes in DNA methylation in the embryonic, extraembryonic and germ cell lineages during mouse embryo development. Development (Camb) 1987;99:371–82.
5. Meehan R, Lewis J, Cross S, Nan X, Jeppesen P, Bird A. Transcriptional repression by methylation of CpG. J Cell Sci Suppl 1992;16:9–14.
6. Barlow DP. Methylation and imprinting: from host defense to gene regulation? Science 1993;260:309–10.
7. Benoit G, Trasler JM. Developmental expression of DNA methyltransferase messenger ribonucleic acid, protein and enzyme activity in the mouse testis. Biol Reprod 1994;50:1312–9.
8. Jue K, Benoit G, Alcivar AA, Trasler JM. Developmental and hormonal regulation of DNA methyltransferase in the rat testis. Biol Reprod 1995;52:1364–71.

9. Jue K, Bestor TH, Trasler JM. Regulated synthesis and localization of DNA methyltransferase during spermatogenesis. Biol Reprod 1995;53:561–9.

10. Trasler JM, Alcivar AA, Hake LE, Bestor T, Hecht NB. DNA methyltransferase is developmentally expressed in replicating and non-replicating male germ cells. Nucleic Acids Res 1992;20:2541–5.

11. Doerksen T, Trasler JM. Developmental exposure of male germ cells to 5-azacytidine results in abnormal preimplantation development in rats. Biol Reprod 1996;55:1155–62.

12. Bestor TH, Tycko B. Creation of genomic methylation patterns. Nat Genet 1996; 12:363–6.

13. Iguchi-Ariga SMM, Schaffner W. CpG methylation of the cAMP responsive enhancer/promoter TGACGTCA abolishes specific factor binding as well as transcriptional activation. Genes Dev 1989;3:612–9.

14. Pfeifer G, Steigerwald S, Hansen RS, Gartler SM, Riggs AD. Polymerase-chain reaction aided genomic sequencing of an X chromosome linked CpG island: methylation patterns suggest clonal inheritance, CpG site autonomy, and an explanation of activity state stability. Proc Natl Acad Sci USA 1990;87:8252–6.

15. Li E, Bestor TH, Jaenisch R. Targeted mutation of the DNA methyltransferase gene results in embryonic lethality. Cell 1992;69:915–26.

16. Lei H, Oh SP, Okano M, Juttermann R, Goss KA, Jaenisch R, et al. De novo DNA cytosine methyltransferase activities in mouse embryonic stem cells. Development (Camb) 1996;122:3195–205.

17. Trasler JM, Trasler DG, Bestor T, Li E, Ghibu F. DNA methyltransferase in normal and Dnmt^n/Dnmt^n mouse embryos. Dev Dyn 1996;206:239–47.

18. Beard C, Li E, Jaenisch R. Loss of methylation activates X^{ist} in somatic but not in embryonic cells. Genes Dev 1995;9:2325–34.

19. Li E, Beard C, Jaenisch R. Role for DNA methylation in genomic imprinting. Nature (Lond) 1993;366:362–5.

20. Tucker KL, Beard C, Dausman J, Jackson-Grusby L, Laird PW, Lei H, et al. Germ-line passage is required for establishment of methylation and expression patterns of imprinted but not nonimprinted genes. Genes Dev 1996;10:1008–20.

21. Bestor TH, Coxon A. The pros and cons of DNA methylation. Curr Biol 1993;3:384–6.

22. Jones PA. DNA methylation errors and cancer. Cancer Res 1996;56:2463–7.

23. Laird PW, Jaenisch R. DNA methylation and cancer. Hum Mol Genet 1994;3:1487–95.

24. Bestor T, Laudano A, Mattaliano R, Ingram V. Cloning and sequencing of a cDNA encoding DNA methyltransferase of mouse cells. J Mol Biol 1988;203:971–83.

25. Yen R-W, Vertino PM, Nelkin BD, Yu JJ, El-Deiry W, Cumaraswarny A, et al. Isolation and characterization of the cDNA encoding human DNA methyltransferase. Nucleic Acids Res 1992;9:2287–91.

26. Bestor TH. Activation of mammalian DNA methyltransferase by cleavage of a Zn-binding regulatory domain. EMBO J 1992;11:2611–8.

27. Bestor TH, Ingram VM. Two DNA methyltransferases from murine erythroleukemia cells: purification, sequence specificity, and mode of interaction with DNA. Proc Natl Acad Sci USA 1983;80:5559–63.

28. Singer-Sam J, Robinson MO, Bellvé AR, Simon MI, Riggs AD. Measurement by quantitative PCR of changes in HPRT, PGK-1, PGK-2, APRT, MTase, and Zfy gene transcripts during mouse spermatogenesis. Nucleic Acids Res 1990;18:1255–9.

29. Numata M, Ono T, Iseki S. Expression and localization of the mRNA for DNA (cytosine-5)-methyltransferase in mouse seminiferous tubules. J Histochem Cytochem 1994;42:1271–6.
30. Leonhardt HL, Page AW, Weier H-U, Bestor TH. A targeting sequence directs DNA methyltransferase to sites of DNA replication in mammalian nuclei. Cell 1992;71:865–73.
31. Carlson LL, Page AW, Bestor TH. Subcellular localization and properties of DNA methyltransferase in preimplantation mouse embryos: implications for genomic imprinting. Genes Devel 1992;6:2536–41.
32. Tucker KL, Talbot D, Lee MA, Leonhardt H, Jaenisch R. Complementation of methylation deficiency in embryonic stem cells by a DNA methyltransferase minigene. Proc Natl Acad Sci USA 1996;93:12920–5.
33. Yoder J, Yen R-W, Vertino PM, Bestor TH, Baylin SB. New 5′ regions of the murine and human genes for DNA (cytosine-5)-methyltransferase. J Biol Chem 1996;271:31092–7.
34. Jones PA, Taylor SM. Cellular differentiation, cytidine analogues, and DNA methylation. Cell 1980;20:85–93.
35. Gabbara S, Bhagwat AS. The mechanism of inhibition of DNA (cytosine-5-)-methyltransferases by 5-azacytosine is likely to involve methyl transfer to the inhibitor. Biochem J 1995;307:87–92.
36. Juttermann R, Li E, Jaenisch R. Toxicity of 5-aza-2′-deoxycytidine to mammalian cells is mediated primarily by covalent trapping of DNA methyltransferase rather than DNA demethylation. Proc Natl Acad Sci USA 1994;91:11797–801.
37. Jones PA. Altering gene expression with 5-azacytidine. Cell 1985;40:485–6.
38. Jones PA, Taylor SM, Mohandas T, Shapiro LJ. Cell cycle-specific reactivation of an inactive X-chromosome locus by 5-azadeoxycytidine. Proc Natl Acad Sci USA 1982;79:1215–9.
39. Ley TJ, DeSimone J, Anagnou NP, Keller GH, Humphries RK, Turner PH, et al. 5-Azacytidine selectively increases τ-globin synthesis in a patient with β+ thalassemia. N Engl J Med 1982;307:1469–75.
40. Constantinides PG, Jones PA, Gevers W. Functional striated muscle cells from non-myoblast precursors following 5-azacytidine treatment. Nature (Lond) 1977;267:364–6.
41. Carr BI, Rahbar S, Asmeron Y, Riggs A, Winberg CD. Carcinogenicity and haemoglobin synthesis induction by cytidine analogues. Br J Cancer 1988;57:395–402.
42. Carr BI, Reilly JG, Smith SS, Winberg C, Riggs A. The tumorigenicity of 5-azacytidine in the male Fischer rat. Carcinogenesis 1984;5:1583–90.
43. Clermont Y. Kinetics of spermatogenesis in mammals: seminiferous epithelium cycle and spermatogonial renewal. Physiol Rev 1972;52:198–236.

Part III

Molecular Regulation of Spermatogenesis

9

CREM: A Master-Switch Governing Transcription in Male Germ Cells

Lucia Monaco and Paolo Sassone-Corsi

Several lines of evidence indicate that highly specialized transcriptional mechanisms ensure stringent stage-specific gene expression in the germ cells. Specific checkpoints correspond to the activation of transcription factors; these regulate gene promoters with a restricted pattern of activity, in a germ cell-specific fashion. There is also evidence that general transcription factors may be differentially regulated in germ cells. For example, TBP (TATA-binding protein) accumulates in early haploid germ cells at much higher levels than in any other somatic cell type. It has been calculated that adult spleen and liver cells contain 0.7 and 2.3 molecules of TBP mRNA per haploid genome-equivalent, respectively, while adult testis contain 80–200 molecules of TBP transcript per haploid genome-equivalent (1). In addition to TBP, TFIIB and RNA polymerase II also were found to be overexpressed in the testis (1). These remarkable features are consistent with the potent transcriptional activity that occurs in a coordinated manner during the germ cell differentiation. Here we discuss the characteristics of cAMP-responsive element modulator (CREM) (2), a transcription factor responsive to the cAMP signaling pathway and whose function is crucial for a normal germ cell differentiation program.

Coupling the Adenylate Cyclase Pathway to Transcription

Intracellular levels of cAMP are regulated primarily by adenylyl cyclase. This enzyme is modulated by various extracellular stimuli mediated by receptors and their interaction with G-proteins (3). The binding of a specific ligand to a receptor results in the activation or inhibition of the cAMP-dependent pathway, ultimately affecting the transcriptional regulation of various genes through distinct promoter-responsive sites. Increased cAMP levels directly affect the function of the tetrameric protein kinase A (PKA) complex (3). Binding of cAMP to two PKA regulatory subunits releases the catalytic subunits, enabling them to phosphorylate target proteins. These are translocated from cytoplasmic and Golgi complex anchoring sites and phosphorylate a number of cytoplasmic and nuclear proteins on serines in the context X-Arg-Arg-X-Ser-X (2, 3). A number of isoforms for both the regulatory and catalytic subunits have been identified, suggesting a further level of complexity in this response (3). In the nucleus, the phosphorylation state of transcription factors and related proteins appears to directly modulate their function and thus the expression of cAMP-inducible genes (2). Thus, there is a direct link between the activation of G-coupled membrane receptors and cAMP-responsive element- (CRE-) mediated gene expression (4).

The analysis of regulatory sequences of several genes allowed the identification of promoter elements that mediate the transcriptional response to increased levels of intracellular cAMP (2). A number of sequences have been identified of which the best characterized is the CRE. A consensus CRE site constitutes an 8-bp palindromic sequence (TGACGTCA) (2). Several genes that are regulated by a variety of endocrinological stimuli contain similar sequences in their promoter regions, although at different positions. A comparison of the CRE sequences identified to date shows that the 5'-half of the palindrome, TGACG, is the best conserved, whereas the 3'-TCA motif is less constant (2). The binding-site specificity appears to require 18–20 bp, because the five or so bases flanking the core consensus have been shown to dictate, in some cases, the permissivity of transcriptional activation (2).

The first CRE-binding factor to be characterized was CREB (CRE-binding protein) (5), but subsequently several additional CRE-binding factors have been identified and the corresponding gene cloned. Most of the CRE-binding proteins were identified by screening a variety of cDNA expression libraries with CRE and ATF sites (6, 7). All these proteins belong to the bZip transcription factor class, while outside of the bZip region, sequence homology between these factors is relatively poor. Various different factors of the CREB/ATF family are able to heterodimerize with each other but only in certain combinations. A "dimerization code" exists that seems to be a property of the leucine zipper structure of each factor (2).

CRE-binding proteins may act as both activators and repressors of transcription. The activator proteins CREB, CREMτ, and ATF-1 mediate transcriptional induction on their phosphorylation by PKA. Their expression is constitutive and widely distributed in various tissues in a housekeeping fashion. Among the repressors, the cAMP-inducible interleukin-1β-converting enzyme repressor (ICER) product deserves special mention. It is generated from a cAMP-inducible alternative promoter of the CREM gene (8, 9). The de novo synthesized ICER protein represses its own promoter, generating an autoregulatory loop (Fig. 9.1). Thus, ICER is an early-response CRE-binding factor and is involved in the dynamics of cAMP-responsive transcription (10).

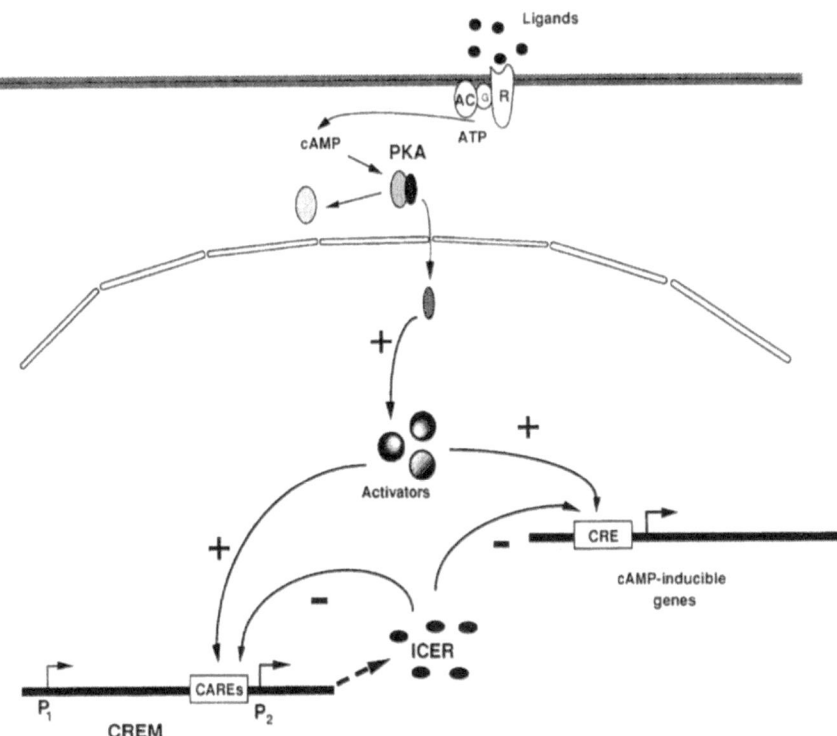

FIGURE 9.1. The cAMP signal transduction pathway: schematic representation of the route whereby ligands at the cell surface interact with membrane receptors (R) and thereby result in altered gene expression. Ligand binding activates coupled G-proteins (G), which in turn stimulates the activity of the membrane-associated adenylyl cyclase (AC). This converts ATP to cAMP, which causes the dissociation of the inactive tetrameric protein kinase A (PKA) complex into the active catalytic subunits and the regulatory subunits. Catalytic subunits migrate into the nucleus where they phosphorylate and thereby activate transcriptional activators such as CREB, CREMτ, and ATF1. These then interact with the cAMP response enhancer element (CRE) found in the promoters of cAMP-responsive genes to activate transcription. Phosphorylated factors also activate transcription from the CREM P_2 promoter via the CARE elements and ultimately lead to a rapid increase in ICER protein levels. ICER represses cAMP-induced transcription, including that from its own promoter. The consequent fall in ICER protein levels eventually leads to a release of repression and permits a new cycle of transcriptional activation.

A Striking Developmental Switch: CREM Expression in the Testis

Expression in the testis provided the first clues that CREM might be distinct from the other bZip factors mediating the cAMP response. CREM was shown to con-

stitute a highly abundant transcript in adult testis; in prepubertal animals, however, CREM is expressed at very low levels (Fig. 9.2). Thus, in the testis CREM is
the subject of a developmental switch in expression (11). Furthermore, charateriz-
ation of the isoforms that were expressed revealed that the abundant CREM
transcript encodes exclusively the CREMτ activator, while in prepubertal testis
only the repressor forms were detected at low levels. Thus, importantly, the

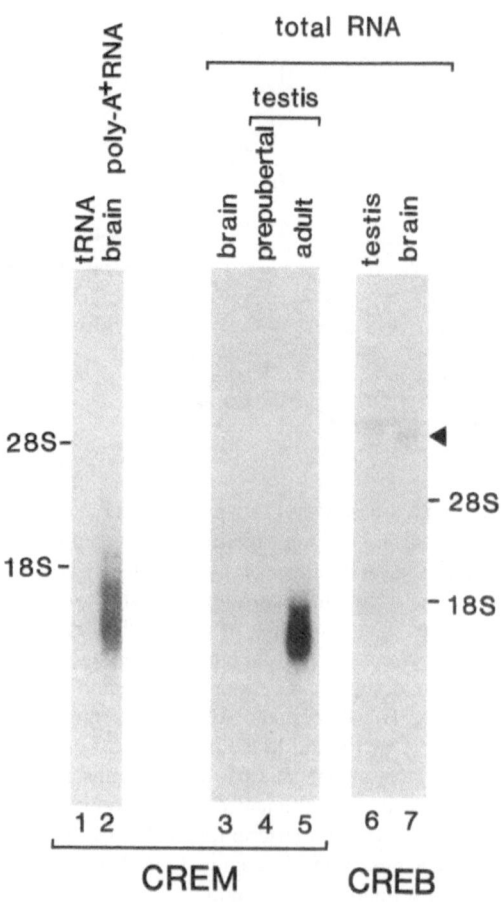

FIGURE 9.2. Developmental-specific expression of CREM in testis shown by Northern blot
analysis of CREM (lanes 1–5) and CREB (lanes 6 and 7) expression in poly-A$^+$ (lane 2)
and total RNA samples (lanes 3–7). Filters containing 5 μg of brain poly-A$^+$ RNA (lane 2)
or 10 μg of total RNA (lanes 3, 4, and 5, respectively) were hybridized with a CREM
cDNA clone. A full-length CREB cDNA fragment was hybridized to a filter containing 10
μg each of testis and brain total RNA (lanes 6 and 7). Arrowheads indicate CREM-specific
bands in brain, the single 2-kb CREM band visible in testis RNA, and the position of the
7-kb CREB transcript.

developmental switch of CREM expression also constitutes a reversal of function (11). To address the precise role played by CREM in the testis, the expression pattern of the RNA and protein have been defined in relation to spermatogenesis (12).

Spermatogenesis corresponds to the sequence of cytological events that results in the formation of haploid spermatozoa from precursor stem cells. This process begins by mitotic division of germ cell spermatogonia to give rise to diploid spermatocytes that themselves replicate their DNA content before undergoing the two successive meiotic divisions, which results in the production of haploid round spermatids. The latter germ cells are sculptured into the shape of the mature spermatozoa in the process of spermiogenesis, which involves an extensive bio-chemical and morphological restructuring. The process occurs in a precise and coordinated manner within the seminiferous tubule. During this entire develop-mental process the germ cells are maintained in intimate contact with the somatic Sertoli cells (13). Indeed, the germ cells are encapsulated within the Sertoli cells, and in this way they are supplied with growth factors and nutrients. Cellular debris generated during spermiogenesis is also processed by the Sertoli cells. As the spermatogonia mature, they move from the periphery toward the lumen of the tubule until the mature spermatozoa are conducted from the lumen to the collect-ing ducts.

Precursor germ cells in a defined segment of the tubule differentiate in syn-chrony, and these zones of differentiation progress along the length of the tubules as a wave (14). A cross-section through a given point of the tubule thus reveals that the proportion of cells at the various stages will change according to its position relative to the waves of differentiation. The cycle of the seminiferous epithelium has been defined as the series of changes occuring in a given area of the seminiferous epithelium between two successive appearances of the same cellular association. There are 12 (I–XII) different cellular associations within the seminiferous epithelium in the mouse (14). During prepubertal development of the testis, the germ cells involved in the first round of spermatogenesis are syn-chronized in their development. Thus, 3–5 days after birth in the mouse, the spermatogonial progenitor cells start to proliferate mitotically; by 9–10 days the first wave of cells have differentiated into preleptotene spermatocytes; at day 13–14 these cells have differentiated into pachytene spermatocytes; by day 18 meiosis is complete and the cells constitute spermatids; finally, by 30–32 days the cells are forming condensing spermatids. The somatic testis cells, namely the Sertoli and Leydig cells, are present at all stages of testicular development. How-ever, the proportion of the total cell population that they constitute decreases as the germ cell population proliferates. Thus, the day of appearance of a gene product during this early stage of development can be used to estimate in which cell type it is expressed.

The finding that the switch in CREM expression does not occur either in the testis of mice irradiated in utero or in the mutant *olt,* but was still intact in the mutant *Tfm,* assigned CREMτ as a germ cell-specific transcript (11). Irradiation in utero is known to selectively kill all germ cells and leave only the somatic cells intact, while the genetic mutants *olt* and *Tfm* have defects in spermatogenesis that arrest germ cell development at the spermatocyte and spermatid stages, respectively. Subsequently, it was demonstrated that the abundant CREMτ transcript appears by day 14. This, combined with analysis of CREMτ transcript expression in fractionated germ cells and in situ hybridization analysis, determined that CREMτ constitutes an abundant transcript from the pachytene spermatocyte stage onward (11).

CREM Regulation of Postmeiotic Gene Expression

By immunohistochemistry and Western blot analysis with a CREM-specific antibody, CREMτ protein is not detected in pachytene spermatocytes but in more mature germ cells which have undergone meiosis. Specifically, CREMτ protein is restricted to round spermatids, mainly at stages VII–VIII of seminiferous tubule differentiation. In mice, overall transcription ceases at about stage IX, when transition proteins and protamines replace the histones so as to compact and condense the chromatin (14). Thus, because CREM protein is not detectable in spermatozoa, CREM transactivator function must be restricted to the late phase of transcription before the compaction of the DNA. The absence of CREM protein in the spermatocytes results from a translational delay of CREM mRNA. Translational control is an important regulatory mechanism of gene expression during spermatogenesis. In particular, it has been reported that in the protamine gene mRNAs are accumulated at high levels to be then translated only at a later time, during the differentiation process (15).

The abundance of CREMτ protein begs the question as to the role that this abundant transactivator plays in haploid germ cells. Several genes have been identified that are transcribed at the time of appearance of the CREM protein and which include CRE-like sequences in their promoter regions (2). To date at least four genes, RT7 (12), transition protein-1 (16), angiotensin-converting enzyme (17), and calspermin (18) have been shown to be targets of CREM-mediated transactivation in germ cells. In all these cases there are several lines of evidence directly implicating CREMτ as a tissue- and temporal-specific regulator. CREM binds to the CRE-related sequences in the promoter of these genes and is able to activate their expression in transient transfection assays. In addition, in an in vitro transcription system using a nuclear extract of seminiferous tubules, the addition of a CREM-specific antibody as well as an excess of CRE competitor decreases RT7 transcription (12). These results suggest that by recognizing various CRE sequences, CREM directs the testis-specific activation of numerous haploid-expressed genes. Interestingly, most of the genes activated at the same time as the appearance of CREM encode structural proteins. For example, transition protein and protamine are detectable around day 22 during mouse spermatogenesis, exactly when CREM proteins are synthesized during spermiogenesis (12).

CREB expression in testis has also been reported (19). The levels of CREB transcript however are dramatically less than that of the CREMτ (see Fig. 9.2) (11,

12). Immunohistochemical analysis with CREB antibodies suggested that the distribution of the CREB protein is similar to that of CREMτ. Unfortunately, in these experiments there was no control for the specificity of the CREB antibodies (19). Indeed, it is likely that the authors were detecting the abundant CREM protein because the anti-CREB antibodies were raised against a synthetic peptide correponding to the P-box domain, which is highly conserved between CREB and CREMτ (12, 19). The function of CREB in testis physiology remains unsubstantiated. Indeed, mice in which various forms of the CREB gene have been deleted by homologous recombination display a normal spermatogenesis process. In contrast, mutation of the CREM gene dramatically affects spermiogenesis.

CREM Knockout Mice: Deficient Spermiogenesis and Increased Apoptosis

The timing of appearance of the abundant CREMτ protein in wild type animals strongly implicates CREMτ as a master controller of haploid gene expression. To address the role of CREM in development and in physiological processes we generated mutant mice with a gene disrupted by homologous recombination in mouse embryonic stem cells (20). We constructed a targeting vector containing a CREM genomic fragment in which a portion of the 3'-terminal exon encoding the DNA-binding domain was deleted and replaced by a PGK-neomycin cassette. The selection of the construct was dictated by the need to inactivate all the numerous CREM and ICER isoforms (2). Reduced fertility was observed in the breeding of the heterozygous mice. Comparison of the homozygous CREM-deficient mice with their normal littermates showed no macroscopic physical aberrations or reduction in body weight. Analysis of internal organs revealed no apparent changes in their structure as compared to wild-type mice. However, the testes of the CREM-deficient mice displayed a reduction of 20%–25% in their weight (Fig. 9.3).

Analysis of the seminal fluid of heterozygous mice compared to normal littermates demonstrated a 46% reduction in the overall number of spermatozoa, a 35% decrease in the ratio of motile spermatozoa, and a twofold increase in the number of spermatozoa with aberrant structures. Most of the aberrant spermatozoa were characterized by a kink and bubble-like structure midway along the tail. Strikingly, analysis of the seminal fluid from homozygous CREM-deficient mice revealed a complete absence of spermatozoa. This result demonstrates a dramatic impairment of spermatogenesis in the CREM-deficient mice. The homozygous males are sterile, demonstrating the necessity of a functional CREM transcription factor for male fertility. The homozygous female mice were fertile and displayed apparently normal ovary structure.

To determine the nature of the sperm deficiency in the CREM-deficient mice, we performed a detailed anatomical analysis of the seminiferous epithelium. Tubuli from CREM-deficient mice display a 20%–30% reduced diameter and

A   B

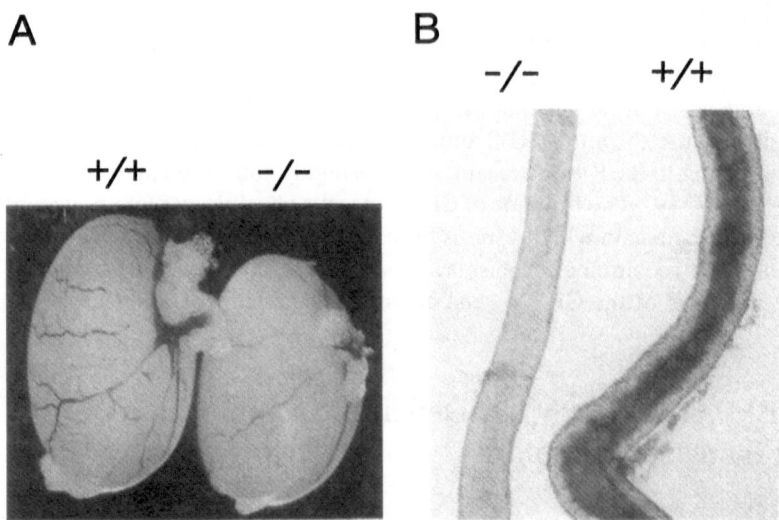

FIGURE 9.3. Aberrant testis phenotype in mice carrying a mutation in the CREM gene. (*A*) Testes from an 8-week old homozygous mutant CREM -/- and a wild-type (+/+) littermate. Note a 20%–25% reduction in size of the mutant testis and reduced vascularization compared to wild-type. (*B*) Photograph of transilluminated seminiferous tubules dissected from the testis of 8-week-old homozygous mutant (-/-) and wild-type (+/+) littermates. The dark section visible in the wild-type tubule that corresponds to regions of spermiation (stages VII–VIII) are completely absent from the homozygous mutant tubules, which are also reduced in diameter.

completely lack the normal spermatogenic wave and the corresponding dark sections (see Fig. 9.3). Squash preparations from consecutive segments of the seminiferous epithelium demonstrate that spermatogenesis in the CREM-deficient mice is interrupted at the stage of very early spermatids. Neither elongating spermatids nor spermatozoa are observed, although somatic Sertoli cells appear to be normal.

The stringent requirement for CREM is manifested by the lack of maturation of the germ cells and by their entering the cell death pathway. Indeed, deletion of CREM causes a 10-fold increase in the number of apoptotic germ cells (Fig. 9.4) (20). In contrast to the Bax-null mice, however, no apoptosis has been observed in other cell lineages of the CREM-deficient mice so far. Thus, the death-interfering function of CREM may be restricted to cells undergoing a meiotic cell cycle.

Hormonal Control of CREM in Testis

The spermatogenic developmental program is under the tight control of the hypothalamus, which regulates the level of pituitary gonadotropins, luteinizing hor-

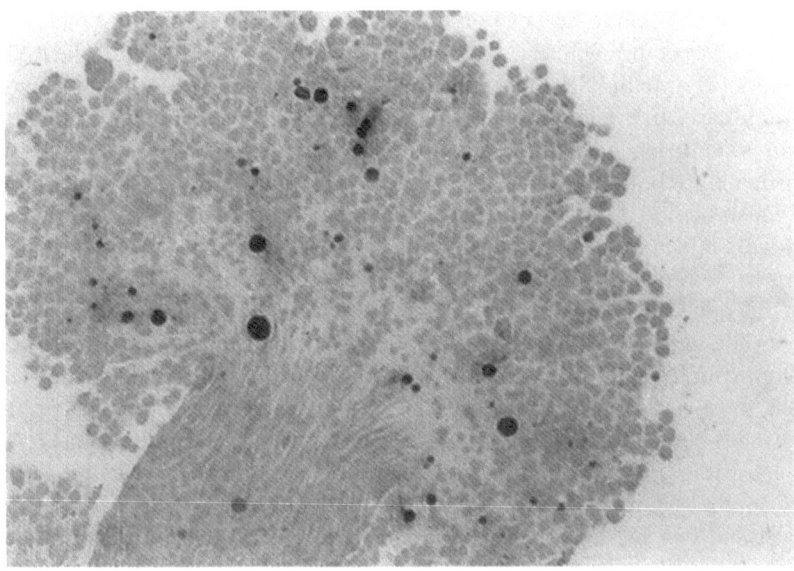

FIGURE 9.4. CREM-deficient mice show increased apoptosis of germ cells: in situ detection of apoptotic cells on squash preparations of seminiferous tubules by labeling with terminal deoxynucleotidyl transferase and digoxigenin-ddUTP. Dark staining bodies represent apoptotic cells whose nuclei contain high concentrations of DNA free ends and correspond to multinucleate giant cells seen in histological sections. The CREM -/- mice contain about 10-fold more apoptotic cells as compared to wild-type animals. The heterozygous mice displayed the same testis phenotype as the wild type.

mone (LH) and follicle-stimulating hormone (FSH). Hormonal stimulation, rather than acting directly on germ cells, is mediated by the Sertoli and Leydig cells. In response to LH stimulation, the Leydig cells, located outside the seminiferous tubules in the interstitial area, produce steroid hormones. FSH and steroid hormones regulate Sertoli cell function depending on the differentiation stage of the tubule. The effects exerted on Sertoli cells by FSH and on Leydig cells by LH are mediated by the cAMP-dependent protein kinase pathway; stimulation of specific adenylyl cyclase-coupled receptors activates the cAMP signal transduction pathway. Although no adenylyl cyclase-coupled receptors have been characterized on germ cells, PKA activity changes during germ cell differentiation to reach a maximum level in spermatozoa. Interestingly, testis-specific isoforms of the catalytic and regulatory subunits of PKA have been described, suggesting a difference in the substrate specificity of the PKA expressed in germ cells compared to the one found in somatic cells. The regulation of CREM function in testis seems to be intricately linked to FSH both at the level of the control of transcript processing and at the level of protein activity.

CREMτ protein is phosphorylated by a cAMP-stimulated PKA activity in round spermatid extracts (12). Phosphorylation occurs at serine 117, the previously described phosphoacceptor site required for transactivation function (2). Thus, CREMτ could represent a developmentally regulated target of the cAMP-dependent PKA, with FSH-induced increases in cAMP levels activating the testis-specific kinase. In this respect, it is noteworthy that FSH-stimulated cAMP production is stage dependent in the seminiferous epithelium (14). At stage VII–VIII a significant FSH-stimulated cAMP production has been reported. Furthermore, it should also be noted that both the PKA RIIα and RIIβ transcripts accumulate at high levels in seminiferous tubules at the same stages as the CREM protein.

Surgical removal of the pituitary gland leads to the loss of CREMτ expression in the rat adult testis (21). Furthermore, in prepubertal animals this treatment prevents the switch in CREM expression at the pachytene spermatocyte stage, thus implicating the pituitary directly in the maintenance of, as well as the switch to, high levels of CREMτ expression. The hormonal stimulus originating from the pituitary has been defined by physiological studies using the golden hamster (21). In the hamster, seasonal variation of gonadal activity is characterized by a cessation of spermatogenesis during winter. This modulation is dependent on the photoperiod and can be experimentally reproduced by artificial lighting. Constant short photoperiod (SP) causes complete gonadal atrophy within 8 weeks. Atrophy is accompanied by a reduction in testosterone and pituitary hormone production. Restoration of long photoperiod (LP) causes a sharp rise in FSH, LH, and prolactin levels and then a progressive increase in testis size and recovery of spermatogenesis. In hamsters maintained under a constant short photoperiod, the testes remain in the atrophied state for up to 12 weeks; however, after this period they spontaneously begin to regenerate. CREMτ expression ceases during the SP-induced gonadal atrophy and then increases upon restoration of long-photoperiod. The absence of CREMτ expression in the atrophied testis coincides with low levels of circulating gonadotrophic hormones. Atrophied testes still contain spermatocytes and early spermatids; thus, CREM expression is not solely under developmental control.

Injections of FSH into hamsters maintained in short photoperiod with fully atrophied testis led to a rapid and significant induction of the CREMτ transcript (21). Indeed, CREMτ expression in the FSH-induced animals is equivalent to that in sexually active adult animals. Nuclear run-on assays confirmed that the induction of CREMτ expression in testis was not associated with an increase in the rate of transcription initiation. Instead, the abundant CREMτ transcript is characterized by the use of an alternative polyadenylation site, which lies only 50 bp downstream of the stop codon (21). The use of this site generates a more stable transcript because 9 of the 10 AUUUA destabilizer elements are excluded (2, 21). CREM is also the first example of a gene whose expression is modulated by a pituitary hormone during spermatogenesis (21). The effect on CREM expression described here could be mediated by yet another hormonal messenger, whose function would be to rapidly convey the FSH-induced signal from Sertoli to germ cells. Indeed, intimate interactions between these cell types have been described (13). An additional facet to the regulation of CREM function in testis comes with the observation that expression of the inducible CREM repressor, ICER, is induced by FSH in Sertoli cells (see following) (22).

One important implication of the FSH function illustrated here is that hormones can regulate gene expression not only at the transcriptional level but also at the level of RNA processing and stability. The wider consequence of these observations is that posttranscriptional events can constitute targets for intracellular signal transduction pathways.

ICER in Sertoli Cells

On the basis of all the data documenting the central role played by CREM during spermatogenesis and the coordinated involvement of gonadotropins, we have further characterized the role FSH on CREM expression in testis. Treatment of primary Sertoli cells with FSH, forskolin, and dbcAMP results in a rapid and transient increase in the PKA-mediated phosphorylation of the activators CREB and CREMτ. Thus, on pituitary hormonal stimulation there is a powerful and prolonged nuclear response to the activation of the adenylyl cyclase pathway. The induced phosphorylation of the activators results in a dramatic stimulation in the expression of the ICER repressor (22). Interestingly, the function of ICER in Sertoli cells seems to be of major physiological importance because one of its targets is the FSH-receptor (FSHR) gene. Specifically, it appears that ICER may be implicated in the downregulation of FSHR expression, leading to the long-term desensitization phenomenon (22). Consistently, no ICER induction is detected in gonadotropin-desensitized Sertoli cells.

The stability of the ICER protein combined with the rapid and transient induction by FSH of ICER RNA are properties consistent with this scenario. ICER-mediated downregulation appears to involve a direct transcriptional repression of the FSHR promoter activity. There is a functional CRE-like site at position -115 in the FSHR promoter that is required for its expression in Sertoli cells (22). On FSH hormonal stimulation, the intracellular levels of ICER increase dramatically. ICER binds efficiently to the CRE-like site in the FSHR promoter, causing the transcriptional downregulation. In this manner, ICER would complete a hormonal negative feedback loop, central to the regulation of Sertoli cell function. Indeed, the kinetics of ICER induction and the potentiation by cychloheximide are again consistent with the model for ICER-negative autoregulation. Thus, ICER appears to be a key determinant in the phenomenon of long-term desensitization of the receptor. Importantly, downregulation of FSHR expression has also been correlated with a decreased stability of the transcript. This observation, together with our results (22), suggest that more than one mechanism may be involved in long-term desensitization of this receptor.

These observations have many implications for the function of ICER in Sertoli cells and more generally in endocrine systems. Namely, they imply that ICER mediates a long-term downregulation of the cAMP transcriptional response following transient hormonal stimulation of the pathway. Thus, it is apparent that the stability of ICER protein may play a fundamental role in defining its function. Indeed, it is apparent that ICER may be involved in the long-term desensitization of other G-protein-coupled receptors in other neuroendocrine tissues (23).

Acknowledgments. We thank all the members of the Sassone-Corsi laboratory and Emiliana Borrelli and Martti Parvinen for help and discussions. This work was supported by grants from the Centre National de la Recherche Scientifique, Institut National de la Santé et de la Recherche Médicale, Centre Hospitalier Universitaire Régional, Fondation de la Recherche Médicale, Rhône-Poulenc Rorer and Association pour Recherche sur le Cancer.

References

1. Sassone-Corsi P. Transcriptional checkpoints determining the fate of male germ cells. Cell 1997;88:163–6.
2. Sassone-Corsi P. Transcription factors responsive to cAMP. Annu Rev Cell Dev Biol 1995;11:355–77.
3. McKnight SG, Clegg CH, Uhler MD, Chrivia JC, Cadd GG, Correll LA, et al. Analysis of the cAMP-dependent protein kinase system using molecular genetic approaches. Recent Prog Horm Res 1988;44:307–35.
4. Montmayeur JP, Borrelli E. Transcription mediated by a cAMP-responsive promoter element is reduced upon activation of dopamine D2 receptors. Proc Natl Acad Sci USA 1991;88:3135–9.
5. Hoeffler JP, Meyer TE, Yun Y, Jameson JL, Habener JF. Cyclic AMP responsive DNA-binding protein: structure based on a cloned placental cDNA. Science 1988;242:1430–3.
6. Hai TY, Liu F, Coukos WJ, Green MR. Transcription factor ATF cDNA clones: an extensive family of leucine zipper proteins able to selectively form DNA binding heterodimers. Genes Dev 1989;3:2083–90.
7. Foulkes NS, Borrelli E, Sassone-Corsi P. CREM gene: use of alternative DNA binding domains generates multiple antagonists of cAMP-induced transcription. Cell 1991; 64:739–49.
8. Molina CA, Foulkes NS, Lalli E, Sassone-Corsi P. Inducibility and negative auto-regulation of CREM: an alternative promoter directs the expression of ICER, an early response repressor. Cell 1993;75:875–86.
9. Stehle JH, Foulkes NS, Molina CA, Simonneaux V, Pévet P, Sassone-Corsi P. Adre-nergic signals direct rhythmic expression of transcriptional repressor CREM in the pineal gland. Nature (Lond) 1993;365:314–20.
10. Lamas M, Lalli E, Foulkes NS, Sassone-Corsi P. Rhythmic transcription and auto-regulatory loops: nuclear pacemaker CREM. Cold Spring Harbor Symp Quant Biol 1996;61:285–94.
11. Foulkes NS, Mellström B, Benusiglio E, Sassone-Corsi P. Developmental switch of CREM function during spermatogenesis: from antagonist to transcriptional activator. Nature (Lond) 1992;355:80–4.
12. Delmas V, van der Hoorn F, Mellström B, Jégou B, Sassone-Corsi P. Induction of CREM activator proteins in spermatids: down-stream targets and implications for haploid germ cell differentiation. Mol Endocrinol 1993;7:1502–14.

13. Skinner MK. Cell-cell interactions in the testis. Endocr Rev 1991;12:45–77.
14. Parvinen M. Regulation of the seminiferous epithelium. Endoc Rev 1992;3:404–17.
15. Kleene KC, Distel RJ, Hecht NB. Translational regulation and deadenylation of a protamine mRNA during spermiogenesis in the mouse. Dev Biol 1989;105:71–9.
16. Kistler M, Sassone-Corsi P, Kistler SW. Identification of a functional cAMP response element in the 5'-flanking region of the gene for transition protein 1 (TP1), a basic chromosomal protein of mammalian spermatids. Biol Reprod 1994;51:1322–9.
17. Zhou Y, Sun Z, Means AR, Sassone-Corsi P, Bernstein KE. cAMP-response element modulator τ is a positive regulator of testis angiotensin converting enzyme transcription. Proc Natl Acad Sci USA 1996;93:12262–6.
18. Sun Z, Sassone-Corsi P, Means A. Calspermin gene transcription is regulated by two cyclic AMP response elements contained in an alternative promoter in the calmodulin kinase IV gene. Mol Cell Biol 1995;15:561–71.
19. Waeber G, Habener J. Novel testis germ-cell specific transcript of the CREB gene contains an alternatively spliced exon with multiple in-frame stop codons. Endocrinology 1992;131:2010–5.
20. Nantel F, Monaco L, Foulkes NS, Masquilier D, LeMeur M, Henriksen K, et al. Spermiogenesis deficiency and germ cell apoptosis in CREM-mutant mice. Nature (Lond) 1996;380:159–62.
21. Foulkes NS, Schlotter F, Pévet P, Sassone-Corsi P. Pituitary hormone FSH directs the CREM functional switch during spermatogenesis. Nature (Lond) 1993;362:264–7.
22. Monaco L, Foulkes NS, Sassone-Corsi P. Pituitary follicle-stimulating hormone (FSH) induces CREM gene expression in Sertoli cells: involvement in long-term desensitization of the FSH receptor. Proc Natl Acad Sci USA 1995;92:10673–7.
23. Lalli E, Sassone-Corsi P. Long-term desensitization of the TSH receptor involves TSH-directed induction of CREM in the thyroid gland. Proc Natl Acad Sci USA 1995;92:9633–7

10

Expression of Activator and Repressor Isoforms of Transcription Factor CREB During Spermatogenesis

Joel F. Habener, Philip B. Daniel, and William H. Walker

The maturation of the germinal cells of the seminiferous epithelium is critically dependent on functional Sertoli and Leydig cells (1). The steroidogenic functions of these cells are regulated by hypothalamic and pituitary gonadotropic hormones. Hypothalamic gonadotropin-releasing hormone (GnRH) stimulates the production of pituitary follicle-stimulating hormone (FSH) and luteinizing hormone (LH), which induce cAMP accumulation in Sertoli and Leydig cells via the activation of adenylate cyclase-coupled receptors for FSH and LH, respectively. The Sertoli cells of the adult testis are somatic postmitotic cells that reach maximum numbers by 2 weeks after birth. They influence the differentiation and metabolism of the germ cells and secrete factors important in spermatogenesis (2).

Spermatogenesis is a cyclical process by which the germinal cells undergo a series of complex maturation steps to become fully differentiated spermatozoa with a haploid genetic complement. The cycle of the seminiferous epithelium has been defined as that series of changes occurring in a given area of the seminiferous epithelium between two successive appearances of the same cellular associations. Leblond and Clermont (3) and Perey et al. (4) have so described 14 (I–XIV) different cellular association stages in the seminiferous epithelium of the rat. The whole process of spermatogenesis in the rat takes approximately 48 days and is divided into almost precisely four 12-day cellular association cycles.

Stimulation of adenylate cyclase by FSH or LH activates cAMP-dependent protein kinase-A (PKA), which, in turn, phosphorylates various substrates responsible for gene regulation (5, 6). Two of these substrates phosphorylated by PKA are the cAMP response element-binding protein (CREB) and the cAMP-responsive element modulator (CREM) (7). Both CREB and CREM are cyclically expressed at high levels during spermatogenesis. Cyclical expression of CREB and CREM in germ and somatic Sertoli cells correlates with the fluctuations in cAMP signaling induced by the pituitary gonadotropic hormones FSH and LH both during sexual maturation of the testis and during the 12-day cycles of sper-matogenesis that occur in the adult testis (8). CREB and CREM are expressed at different times during the spermatogenic cycle, undergo programmed sequential switches from activator to repressor isoforms by mechanisms of alternative exon splicing and promoter usage, and are autoregulated by cAMP signaling in oppos-ing directions (9). cAMP response elements located in the promoter of the CREB gene upregulate the expression of activator CREBs, whereas cAMP autoregula-tory response elements in the internal promoter of the CREM gene induce expres-sion of repressor CREM isoforms (9, 10). Here, we discuss our views of the regulation of the expression of the cAMP-responsive transcription factors, CREB and CREM, in spermatogenesis.

Expression of CREB and CREM

The activation of gene transcription by the cAMP response element-binding pro-tein (CREB) represents a final step of the cAMP-signaling pathway. CREB is a member of the bZIP family of transcription factors that consist of two functionally distinct domains: a carboxy-terminal dimerization and DNA-binding (bZIP) do-main and an amino-terminal transcriptional transactivation domain (6, 7). The transactivation of gene transcription by CREB is dependent on the phosphoryla-tion of a single serine within the P-box (or KID) by cAMP-dependent protein kinase A (PKA) (7). The gene encoding CREB consists of at least 12 exons, several of which are alternatively spliced, resulting in the synthesis of a variety of CREB isoforms (Fig. 10.1). Alterations in the exonic complement of CREB has been shown to affect the regulatory properties of CREB, as splicing in of exon D in CREB results in an enhancement of transactivation activity (7), whereas splic-ing in of exon W attenuates transactivation activity (11, 12).

CREB is expressed in nearly all tissues tested thus far. However, in the testis, novel isoforms of CREB arise from alternative splicing of exons, resulting in the synthesis of mRNAs that alter the protein-coding regions of CREB (11, 13, 14). An unusual property of these alternatively spliced exons, designated exons W, Y, and ψ, is that when spliced into the CREB mRNA all translational reading frames are blocked so as to terminate translation, resulting in the synthesis of CREB isoforms lacking the bZIP DNA-binding domain and nuclear translocation signal. In the rat testis, these alternatively spliced mRNAs encoding isoforms of CREB are cyclically expressed during development of the germ and Sertoli cells (9).

It has been shown that the inclusion of the alternatively spliced exon W con-verts the CREB message from a monocistronic to a polycistronic mRNA that encodes novel repressor or inhibitor CREBs (I-CREBs) (12, 15). Exon W contains an in-frame stop codon that terminates translation amino proximal to the DNA-binding bZIP domain of CREB, resulting in the reinitiation of translation at two downstream internal methionine codons (Figs. 10.1 and 10.2). The two internally translated truncated CREB isoforms, I-CREB(l) and I-CREB(s), retain the CREB reading frame and thereby consist of the DNA-binding domain devoid of the

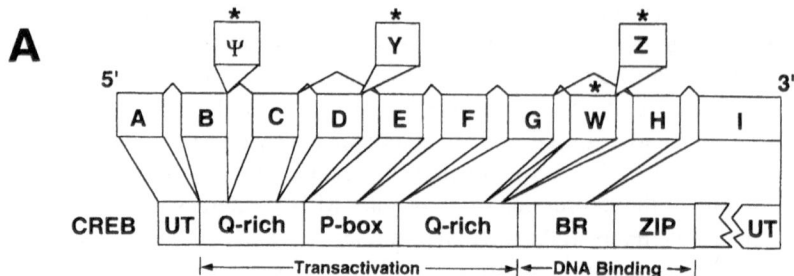

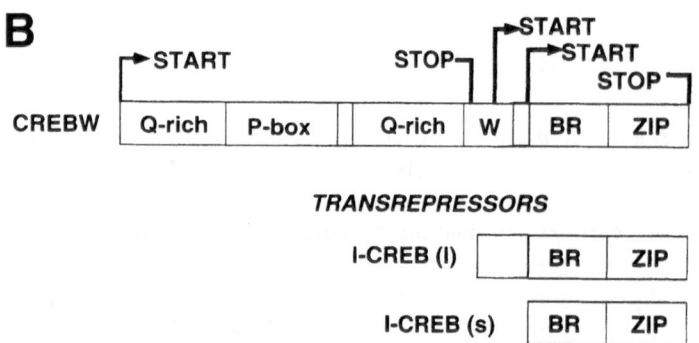

FIGURE 10.1. The exonic organization of the CREB gene, the functional domains of CREB, and the formation of transrepressor I-CREB isoforms by internal translation in response to the alternative splicing of exon W. (*A*) The CREB gene consists of at least 13 exons in which exons ψ, Y, D, W, and Z are alternatively spliced in the testis. Exons ψ, Y, W, and Z all contain in-frame translational stop codons (asterisks). The transactivation region of CREB consists of two domains relatively enriched in the amino acid glutamine (Q-rich) and a domain that is phosphorylated by cAMP-dependent protein kinase A and other protein kinases (P-box). The DNA-binding region consists of a basic region (BR) enriched in lysines and arginines and a dimerization domain (ZIP). UT, untranslated regions. (*B*) The alternative splicing-in of exon W in the CREB W mRNA results in a termination of translation in exon W and a reinitiation of translation in frame on AUG codons located 3 and 25 codons downstream from the stop codon (UAA). The two internally translated proteins designated inhibitor CREBs long and short, I-CREB(l) and I-CREB(s), consist of the DNA-binding domain devoid of the transactivating region and thereby act as a trans-repressors of activator CREBs (14, 15).

amino-proximal transactivation domains. As a consequence, these two CREB isoforms inhibit the binding of activator forms of CREB to cAMP response elements (CREs) and are dominant-negative inhibitors of cAMP-mediated gene transcription (Fig. 10.3). The cyclical alternative splicing of exon W during spermatogenesis in the rat testis is remarkable in that it results in a periodic, alternative

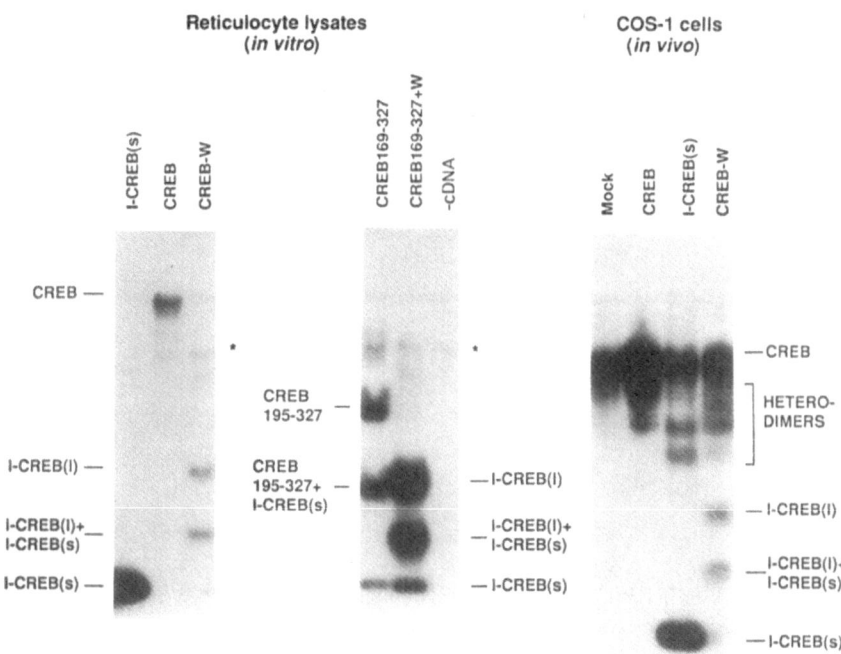

FIGURE 10.2. Internal translation of inhibitor I-CREBs from CREB-W mRNA in which exon W is spliced into the message. I-CREBs were synthesized in vitro in the TNT coupled transcription-translation system in reticulocyte lysates (left) or in vivo in Cos-1 cells transfected with a plasmid that expresses CREB-W mRNAs. In addition to CREB and CREB-W mRNAs, 5'-truncated mRNAs CREB169–327 and CREB169–327+W were synthesized and translated. The translated protein products were analyzed by electrophoretic mobility gel shift using a 20-bp double-stranded oligonucleotide containing a cyclic AMP response element that binds isoforms of CREB. Shown are the various homodimer and heterodimer combinations of CREB, I-CREB(l), and I-CREB(s) that contain the carboxyl-terminal DNA-binding bZIP domain of CREB (14).

expression of CREB isoforms that either activate or repress gene transcription (Fig. 10.4).

In the human testis there exists a novel human-specific CREB mRNA containing an alternatively spliced exon Z (15). This exon Z is spliced in between CREB exons W and H, always in conjunction with the splicing in of exon W. Notably, an in-frame stop codon in exon Z terminates the translation of I-CREB(l) and results in an enhancement of the internal translation of the I-CREB(s) isoform initiating downstream of exon Z. Sequencing of rat, mouse, and human genomic DNA reveals that exon Z is located within the intron between exons W and H. However, the bases immediately flanking exon Z, required for splicing, are mutated in the mouse and rat DNA so that exon Z remains unspliced (15). These findings suggest

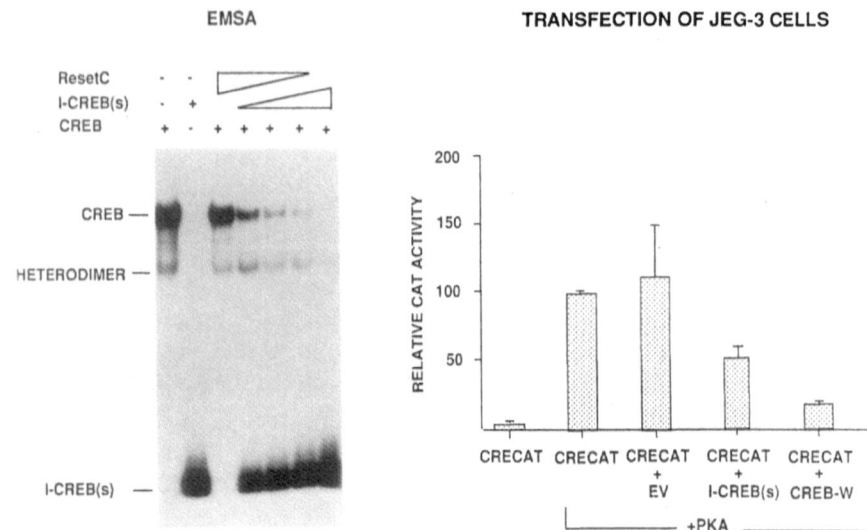

FIGURE 10.3. I-CREB(s) derived by internal translation of CREB-W mRNA inhibits binding of CREB to a cAMP-response element (CRE) and represses PKA-mediated gene transcription. Left, a DNA-binding electromobility shift assay (EMSA) in which increasing amounts of I-CREB(s) compete CREB homodimers and heterodimers of CREB and I-CREB(s) from binding to a ^{32}P-labeled oligonucleotide containing a CRE; right, a transfection-expression-transcription assay in which plasmids expressing CREB, CREB-W, or I-CREB(s) were transfected and expressed in placental JEG-3 cells in combination with a transcriptional reporter plasmid consisting of a CRE linked to the minimal promoter of the glycoprotein hormone α-subunit gene and the gene encoding bacterial chloramphenicol acetyl transferase (CRECAT). PKA, cAMP-dependent protein kinase A expression plasmid co-transfected with the described plasmids; EV, empty vector control plasmid (14).

the evolution of a unique human-specific alternative splicing event that suppresses the expression of one, I-CREB(l), but not the other, I-CREB(s), dominant-negative inhibitor of CREB.

The levels of CREB mRNA increase in cell association stages II–VI following increases in FSH-induced cAMP levels in stages XII–V. Levels of CREB mRNA then fall rapidly to near-undetectable levels in stages VII–XIV as cAMP concentrations decrease. Characterization of the CREB promoter identified three CREs responsible for cAMP induction of transcription (6, 16). Phosphorylation by cAMP-dependent PKA activates CREB bound to the CREs on the CREB promoter, thereby stimulating transcription and the production of additional CREB, resulting in an autopositive feedback loop. This autopositive regulation of CREB gene expression is proposed to account for the rapid and large stage-specific increase in CREB mRNA that accumulates in the nuclei of Sertoli cells during stages II–VI (6, 9, 16).

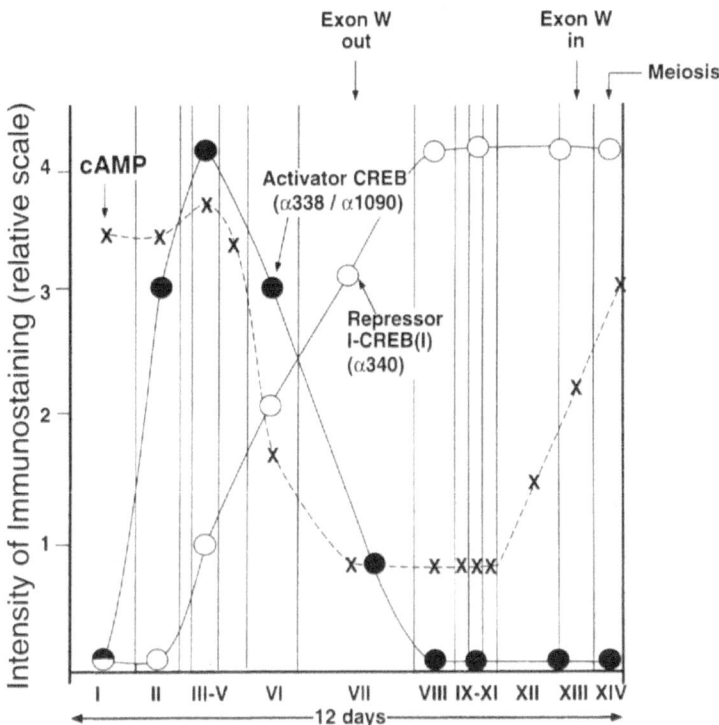

FIGURE 10.4. Cyclical expression of activator and repressor I-CREB(1) during spermatogenesis showing the relative intensity of immunostaining of germ cells in seminiferous tubules at various stages of the 12-day spermatogenic cycle. Immunostaining was done with antisera α338 and α1090 specific for CREB (α338 does not cross-react with CREM) and α340 raised to a synthetic peptide that contains epitopes unique to the translational open reading frame in exon W that forms the N-terminal sequence of I-CREB(1). In separate experiments employing the reverse transcriptase-polymerase chain reaction (RT-PCR) to select CREB mRNAs in sequential segments of isolated tubules, the times of splicing-in and -out of exon W were determined (arrows) (splice cycling experiments, unpublished). Also shown on the plot are the levels of cAMP that fluctuate during the spermatogenic cycle as determined by Rannikko et al. (8). Cell association stages of the tubules are shown on the abscissa (Roman numerals I–XIV), as well as the occurrence of meiosis at stage XIV.

The formation of inhibitor CREB isoforms by the inclusion of exon W in the RNA and the synthesis of I-CREB repressors precedes a pronounced fall in the levels of CREB mRNA, suggesting that I-CREBs antagonize the synthesis of CREB mRNA (2). The formation of I-CREBs as well as the decreased production of full-length activator CREB incurred by the splicing of exon W is predicted to interrupt a positive feedback loop (Fig. 10.5).

Notably, transcription of the CREM gene is also autoregulated by cAMP sig-

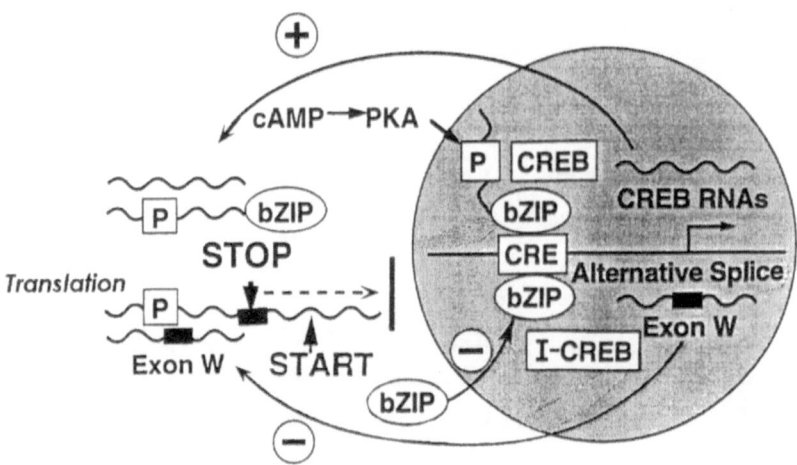

FIGURE 10.5. Model of hypothetical cyclical autoregulation of the expression of the CREB gene during spermatogenesis. Alternative splicing of exon W in CREB mRNA during the spermatogenic cycle switches the synthesis of CREB from an activator CREB isoform (CREB) to a repressor CREB isoform (I-CREB) containing the DNA-binding domain (bZIP) and devoid of the transactivation domain (P). Exon W contains in-frame translational termination codons, stops translation (Stop), and allows for new downstream internal translation (START) encoding repressor I-CREBs. cAMP activates protein kinase A that phosphorylates CREB and activates transcription of the CREB gene which contains three cAMP response elements in its promoter.

naling although via a distinctly contrasting mechanism. Although the promoter at the 5'-end of the CREM gene (P1) that programs transcripts encoding activator CREM isoforms is reportedly not inducible, cAMP activates an internal promoter (P2) located in the 3'-region of the CREM gene, resulting in the synthesis of repressor CREM isoforms (10). cAMP-responsive CREB and CREB-like activator proteins interact with four CREs (cAMP autoregulatory response elements) in the P2 promoter to stimulate transcription (10). The mRNA that encodes the interleukin-1β-converting enzyme repressor (ICER) repressor consists of a short ICER-specific region followed by the bZIP DNA-binding domain.

Although the autopositive upregulation of CREB gene expression in male germ cells is interrupted by the switch in expression from activator to inhibitor CREBs (I-CREBs), the I-CREBs appear to be expressed at low levels in the somatic Sertoli cells, leaving unanswered the mechanisms by which the cyclical upregulation of CREB is interrupted in Sertoli cells. It was reported that the ICER repressor is expressed in primary rat Sertoli cells in response to FSH and is proposed to downregulate the transcription of the FSH receptor gene in these cells (17). We have observed that FSH-induced expression of ICER downregulates the expression of the CREB gene in primary rat Sertoli cells and thereby provides a potential explanation for the cyclical fluctuations in CREB gene expression during spermatogenesis (Walker et al., unpublished observations). Further, we found by studies of isolated rat seminiferous tubules that CREB and ICER are expressed at different stages of the 12-day cycle of spermatogenesis. Because ICER is a candidate repressor of the CREB gene, it may be an important factor responsible for the delayed induction of the CREB gene by FSH and cAMP in Sertoli cells. Together, the two regulators, CREB and ICER, may be responsible for regulating the expression of cAMP-inducible genes in Sertoli cells to specific stages of germ cell development.

Summary

The CREB and CREM genes are expressed cyclically at particularly high levels and at different stages of germ cell development in the testis. cAMP-dependent autoregulation of the transcription of the genes and alternative splicing of exons, resulting in the interconversion of transactivator to transrepressor isoforms, play an important role in spermatogenesis during both sexual maturation and in the repeated cycles of germ cell development that take place in the testis of adult animals. Many uncertainties remain regarding the role of CREB and CREM in spermatogenesis. Future studies should be directed toward obtaining a better understanding of the stage-specific target genes that are regulated by CREB and CREM as well as by other, to date unidentified, cAMP-responsive transcription factors. Given the preemptive role of cAMP signaling in spermatogenesis, it seems reasonable to anticipate that the alternative splicing of the CREB and CREM RNAs is regulated by cAMP. It is tempting to speculate that the promoters of genes encoding specific spliceosome components contain cAMP-responsive sequences regulated by CREB and CREM.

A final enigma remains in the interrelationships among the expression of the CREB and CREM genes in Sertoli and germ cells. It is worth noting that the majority of cAMP regulation in spermatocytes is most likely controlled by CREB/I-CREB ratios because activator forms of CREM are not detected until later stages of germ cell differentiation (9). During critical times of early germ cell development, alterations in splicing protocols for exon W, perhaps mediated by hormonal signals (cAMP), would dictate relative levels of repressor I-CREBs and activator CREBs. Competition between CREB and I-CREBs would then determine the rates of transcription of cAMP-regulated genes including the CREB gene.

Acknowledgments. We thank T. Budde for help in preparation of the manuscript. This work was supported in part by USPHS grant DK25532. J.F.H. is an investigator with the Howard Hughes Medical Institute.

References

 1. Lostroh AJ. Hormonal control of spermatogenesis. In: Spilman CH, Lobl TJ, Kirton KT, eds. Regulation mechanisms of male reproductive physiology. Amsterdam: Excerpta Medica, 1976;13–23.

2. Kissinger C, Skinner MK, Griswold MD. Analysis of Sertoli cell-secreted proteins by two-dimensional gel electrophoresis. Biol Reprod 1982;27:233–40.
3. Leblond CP, Clermont Y. Definition of the stages of the cycle of the seminiferous epithelium in the rat. Ann NY Acad Sci 1952;55:548–70.
4. Perey B, Clermont Y, LeBlond CP. The wave of the seminiferous epithelium in the rat. Am J Anat 1961;108–9:47–76.
5. Habener JF. The cyclic AMP second messenger signaling pathway. In: de Groot, ed. Endocrinology. Philadelphia: Saunders, 1995;77–92.
6. Walker WH, Fucci L, Habener JF. Expression of the gene encoding transcription factor cyclic adenosine 3',5'-monophosphate (cAMP) response element-binding protein (CREB): regulation by follicle stimulating hormone-induced cAMP signaling in primary rat Sertoli cells. Endocrinology 1995;136:3534–5.
7. Habener JF, Miller CP, Vallejo M. Cyclic AMP-dependent regulation of gene transcription by CREB and CREM. In: Litwack G, ed. Vitamins and hormones. San Diego: Academic Press, 1995;51:1–57.
8. Rannikko A, Penttila T-L, Zhang F-P, Toppari J, Parvinen M, Huhtaniemi I. Stage-specific expression of the FSH receptor gene in the prepubertal and adult rat seminiferous epithelium. J Endocrinol 1996;151:29–35.
9. Walker WH, Habener JF. Role of transcription factors CREB and CREM in cAMP-regulated transcription during spermatogenesis. Trends Endocrinol Metab 1996;7:133–8.
10. Foulkes NS, Sassone-Corsi P. More is better: activators and repressors from the same gene. Cell 1992;68:411–4.
11. Waeber G, Meyer TE, LeSieur M, Hermann HL, Gérard N, Habener JF. Developmental stage-specific expression of cyclic adenosine 3',5'-monophosphate response element-binding protein CREB during spermatogenesis involves alternative exon splicing. Mol Endocrinol 1991;5:1418–30.
12. Walker WH, Girardet C, Habener JF. Alternative exon splicing controls a translational switch from activator to repressor isoforms of transcription factor CREB during spermatogenesis. J Biol Chem 1996;271:20145–50.
13. Waeber G, Habener JF. Novel testis germ cell-specific transcript of the CREB gene contains an alternatively spliced exon with multiple in-frame stop codons. Endocrinology 1992;131:2010–5.
14. Ruppert S, Cole TJ, Boshart M, Schmid E, Schülz G. Multiple mRNA isoforms of the transcription activator protein CREB: generation by alternative splicing and specific expression in primary spermatocytes. EMBO J 1992;11:1503–12.
15. Girardet C, Walker WH, Habener JF. An alternatively spliced polycistronic mRNA encoding cyclic adenosine 3',5'-monophosphate (cAMP)-responsive transcription factor CREB (cAMP response element binding protein) in human testis extinguishes expression of an internally translated inhibitor CREB isoform. Mol Endocrinol 1996;10:879–91.
16. Meyer TE, Waeber G, Lin J, Beckmann W, Habener JF. The promoter of the gene encoding 3',5'-cyclic adenosine monophosphate (cAMP) response element-binding protein contains cAMP response elements: evidence for positive autoregulation of gene transcription. Endocrinology 1993;132:770–80.
17. Monaco L, Foulkes NS, Sassone-Corsi P. Pituitary follicle-stimulating hormone (FSH) induces CREM gene expression in Sertoli cells: involvement in long-term desensitization of the FSH receptor. Proc Natl Acad Sci USA 1995;92:10673–7.

11

Repression and Activation of Protamine mRNA Translation During Murine Spermatogenesis

Robert E. Braun

Translational Regulation During Murine Spermatogenesis

Murine spermatogenesis initiates a few days after birth and continues for the duration of the sexual life of the animal. Spermatogenesis takes approximately 35 days and consists of the mitotic proliferation of spermatogonial cells, meiosis, and spermiogenesis, the haploid spermatid differentiation stage (Fig. 11.1). Transcription is ongoing throughout spermatogonial proliferation, meiosis, and the early spermiogenesis. In fact, unlike spermatogenesis in *Drosophila* where there is no postmeiotic transcription, in the mouse there is considerable elevation in the transcriptional apparatus shortly after meiosis (1). Transcription continues until the transition from the round to the elongating spermatid, and then ceases several days before to the completion of spermiogenesis (2, 3). The study of transcriptional activity has relied primarily on metabolic labeling studies, which are limited in their sensitivity. The available data do not allow us to distinguish between transcriptional silencing at the transition from round spermatid to elongating spermatid (about step 9), or the elongating spermatid to elongated spermatid transition (about step 13). Distinguishing between these two possibilities is important if one is to understand the reason for transcriptional silencing. Cessation of transcription at step 9 would likely involve developmental changes in the transcriptional apparatus, whereas transcriptional arrest at step 13 could be attributed to changes in chromatin structure that occur as chromosome condensation commences.

Chromosome condensation in spermatogenesis is initiated by the transition proteins (TPs) and is completed by the protamines (4). Both of these classes of genes are under translational control. In the case of the two protamine genes, Prm-1 and Prm-2, the genes are first transcribed in step 7 spermatids, but their respective mRNAs are not translated until approximately step 13 (5–8). Transla-

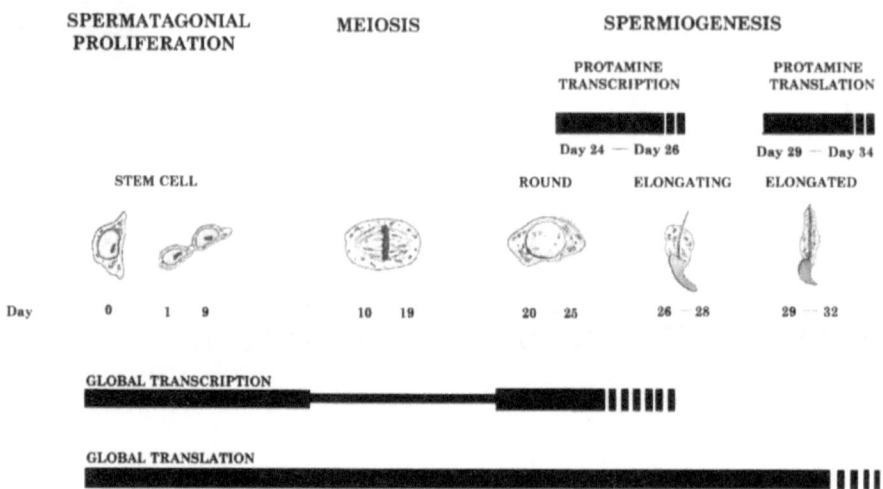

FIGURE 11.1. Schematic of spermatogenesis in the prepubertal mouse.

tional repression of Prm-1 and Prm-2 occurs at a time when other mRNAs are actively translated. Thus, repression of protamine translation is specific. In addition, other mRNAs are translationally repressed during meiosis and in early round spermatids, but are translated in late round or elongating spermatids (9, 10). It is not evident why mRNAs that are translated before the cessation of transcription are under translational control, nor are the mechanisms controlling their translation necessarily similar to that of the transition proteins and the protamines. Elucidation of the mechanism of translational repression in round and elongating spermatids, and the mechanism of translational activation in elongated spermatids, is one of our laboratory's primary goals.

The 3'-Untranslated Region of the Protamine mRNA

Translational repression of Prm-1 mRNA is controlled by sequences in its 3'-untranslated region (3'-UTR). Demonstration that the Prm-1 3'-UTR controls its translation was shown using gene fusions in transgenic mice (11). Chimeric mRNAs containing the human growth hormone coding sequences fused to the Prm-1 3'-UTR are translationally repressed in round and elongating spermatids, and translationally active in elongated spermatids, in the same manner as the endogenous Prm-1 mRNA. Thus, sequences in the 3'-UTR mediate both proper translational repression in round spermatids and translational activation in elongated spermatids.

Translational inhibition of Prm-1 mRNA is absolutely essential for the completion of spermiogenesis (12). Premature translation of a Prm-1 transgene results in precocious chromosome condensation and dominant male sterility. This dominant male sterility is the result of the sharing of postmeiotic gene products between spermatids (13). Sharing of gene products is possible because of the intercellular bridges that connect cells within developing clones of spermatids (14). Overexpression of Prm-1 protein in step 13 spermatids does not affect spermiogenesis (15, 16); thus, overexpression per se is not responsible for the phenotypes observed with premature synthesis of Prm-1.

Premature translation of Prm-1 mRNA leads to two clearly distinguishable histological phenotypes. As detected by light microscopy, the majority of the sterile males contain late-stage spermatid defects characterized by abnormal head morphogenesis and a failure to complete the final stages of spermiogenesis. A second phenotype, seen in one founder transgenic animal and all males in one transgenic line, was a complete developmental arrest at the round spermatid stage observed in every seminiferous tubule examined.

When Prm-1 protein is synthesized prematurely, it is localized to the nucleus of round spermatids. Because Prm-1 is a highly basic protein, it is possible that the phenotype associated with premature translation of Prm-1 mRNA is a result of a general inhibition of transcription. However, comparison of the accumulation of the endogenous Prm-1 and Prm-2 mRNAs relative to actin suggests that global transcription is not affected. Nonetheless, one can envision that transcription of one or more essential genes is reduced or blocked as a consequence of premature Prm-1 synthesis, resulting in spermiogenic arrest. Alternatively, it is also conceivable that spermatids have a developmental checkpoint that monitors the events associated with nuclear condensation, and that it triggers an arrest in spermiogenesis if nuclear condensation is aberrant.

The effect of premature synthesis of Prm-1 on the synthesis of the transition proteins (TPs) and the endogenous protamines was also studied. To analyze the synthesis of the various testis basic proteins, total testis basic proteins were isolated from sonication-sensitive and sonication-resistant nuclei and separated by acid-urea gel electrophoresis. Spermatid nuclei normally become sonication-resistant in wild-type mice after deposition of the TPs and protamines (17). In animals with the late-stage spermatid-defect phenotype, we detected the presence of Prm-1, transition protein 2 (TP2), and the precursor of Prm-2, only in sonication-resistant spermatid nuclei. The absence of Prm-1 protein in sonication-sensitive nuclei suggests that accumulation of Prm-1 leads to sonication resistance. Prm-2 is normally synthesized as a precursor protein of 106 amino acids and is processed through a series of proteolytic cleavages to a final size of 63 amino acids (18, 19). Interestingly, in animals that displayed the late-stage spermatid-defect phenotype, only the unprocessed form of Prm-2 was detected. Thus, in mice with the late-stage spermatid defect, the transition from nucleo-histones to protamines is incomplete. The failure to properly process the Prm-2 precursor protein suggests that the processing of Prm-2 is coupled with its normal co-deposition with Prm-1 protein. The deposition of Prm-1 protein on chromatin before synthesis of the TPs demonstrates that removal of the TPs is not required for Prm-1 deposition. However, our data suggest that the deposition of Prm-1 protein is abnormal and that it interferes with the normal events of spermiogenesis. In summary, these data suggest that the failure to complete spermatid differentiation is caused by a failure to complete the transition from nucleo-histones to nucleoprotamines and demonstrate that chromatin condensation depends on the correct temporal synthesis of the TPs and the protamines.

Protamine RNA-Binding Proteins

Translational repression by sequences in the 3′-UTR of the Prm-1 mRNA necessitates communication between the two ends of the message. Several models can be considered. Sequences in the 3′-UTR could function to localize the mRNA to a subcellular compartment within the cell that is devoid of ribosomes. In the case of Prm-1 mRNA, there is no evidence of compartmentalization of the mRNA in the cytoplasm of round and elongating spermatids (20). Alternatively, proteins or antisense RNAs could bind to the Prm-1 3′-UTR and to a region in the 5′-end of the message, for example, the ^{7}M-GpppN CAP, and interfere with the assembly of the 43S ribosomal subunit on the end of the message. Factors bound to the 3′-UTR could also initiate a process of complete mRNA masking by the same protein, or other proteins, thus resulting in a translationally compromised ribonucleoprotein particle. Finally, proteins bound to the 3′-UTR could interfere with the binding of poly(A)-binding protein to the poly(A) tail. Recent evidence in yeast suggest that poly(A)-binding protein can physically couple the two ends of the message and that this coupling is important for translation initiation (21, 22).

Determining if any of these models function for Prm-1 mRNA requires the identification of the factors that bind to the 3′-UTR. We have previously described the biochemical characterization of a testis-specific RNA-binding protein that interacts with the 3′-UTRs of the Prm-1 and Prm-2 mRNAs (23). RNA electrophoretic mobility shift assays (EMSA) with radiolabeled Prm-1 3′-UTR and testis protein extracts detect a closely spaced doublet on native polyacrylamide gels. Cross-linking studies with ultraviolet light indicate that the protein components of the complex are approximately 48 and 50 kDa. The binding activity is present in the cytoplasm of pachytene spermatocytes and round spermatids and is barely detectable in elongated spermatids. Competition assays reveal that the complex is specific for Prm-1 and Prm-2 mRNAs. Mutational analysis and RNAse footprinting studies have been used to map the binding site to a specific region in the 3′-UTRs. Comparison of the binding sites for Prm-1 and Prm-2 show that there is a 9-nt sequence that is identical at 7 of the 9 nts between Prm-1 and Prm-2, and mutation of the conserved 7 nts completely abolishes protein binding (Fajardo et al., unpublished data).

The binding specificity of the 48/50-kDa proteins, coupled with their spatial and temporal localization, are properties one would expect for a translational repressor of Prm-1 mRNA. However, previous transgenic analysis of the cis-acting sequences required for Prm-1-like translational control in vivo indicate that the binding site for the 48/50-kDa complex is not required (Fig. 11.2B). It is possible that, despite the highly suggestive properties of the binding activity, the

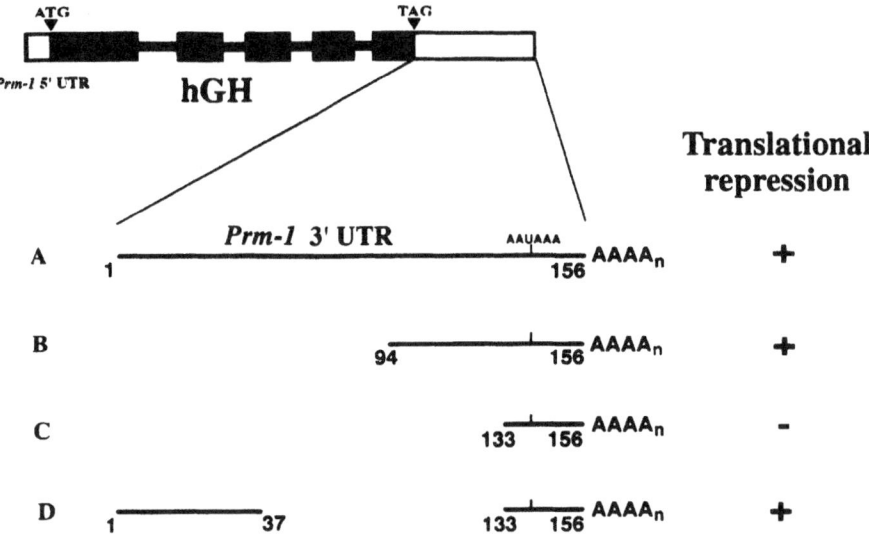

FIGURE 11.2. Summary of chimeric mRNAs assayed for Prm-1-like (protamine gene-like) translational control in transgenic mice. All transgene constructs generate mRNAs with heterologous 5'-UTRs (untranslated regions) of 159 nt, labeled Prm-1 5'-UTR, and represented as an open box at the beginning of the reporter construct. The 5'-UTR used in these studies contains 91 nt of Prm-1 5'-UTR, 7 nt of linker, and 61 nt of hGH (human growth hormone) 5'-UTR. Each transgenic mRNA contains the complete hGH-coding sequence and introns represented as thick and thin solid black boxes, with translational start and stop codons labeled. The full-length Prm-1 3'-UTR, or specific deletion variants of the Prm-1 3'-UTR, were linked to hGH as represented by the open box at the end of the reporter construct. The fusions (A–D) generate 3'-UTRs that contain the first 7 nt of the hGH 3'-UTR, and either (A) the full-length 156-nt Prm-1 3'UTR, (B) the 3'-most 62 nts of the Prm-1 3'-UTR, (C) the 3'-most 23 nts of the Prm-1 3'-UTR, or (D) the first 37 nt of the 5'-end of the Prm-1 3'-UTR fused to the 3'-most 23 nts of the Prm-1 3'-UTR. All chimeric mRNAs were evaluated for Prm-1-like temporal translational control using either or both immunocytochemistry and prepubertal Western blot analysis. From Braun et al. (11) (A); Braun (24) (A, C); and Fajardo et al. (38) (D), with permission.

48/50-kDa protein complex is required for a different aspect of Prm-1 metabolism. Alternatively, it is also possible that more than one region of the Prm-1 3'-UTR is sufficient for translational repression in vivo. To test this possibility, we constructed a transgene containing the binding site for the 48/50-kDa protein (Fig. 11.2D). Analysis of several transgenic founder animals and males from two established transgenic lines clearly shows that transgene D is translationally repressed in round and elongating spermatids (Fajardo et al., in manuscript). Northern blotting shows that the transgenic mRNA is detectable in the prepubertal testes of 26-day-old mice, yet the reporter protein, human growth hormone (hGH), is not detectable until 32 days after birth. This pattern of RNA and protein expression in the prepubertal testis is similar to that of transgenes containing the entire Prm-1 3'-UTR (11). Immunocytochemistry of adult testis sections demonstrates that the hGH protein is present in the cytoplasm of elongated spermatids and, to a slight extent, in the acrosome of round and elongating spermatids. The localization of hGH protein to the developing acrosome is an indicator of premature translation of hGH mRNA in round and elongating spermatids (11). Thus, the binding site for the 48/50-kDa protein appears to inhibit translation in round spermatids, although it may not do so completely.

Immunolocalization of the hGH protein in mice carrying transgenic construct D shows that the mRNA is translated in elongated spermatids, but the hGH signal observed is unusually weak. One possibility for the weakness of the protein signal is that the transgenic mRNA is unstable and that little mRNA is present in elongated spermatids. To test this possibility we performed in situ hybridization. Relative to the levels of the endogenous Prm-1 mRNA, the transgenic mRNA actually appears to be more stable. Thus, mRNA stability does not seem to be the cause of the low protein accumulation in elongated spermatids. A second possibility is that the transgenic mRNA is poorly activated for translation. To test this possibility, we performed Northern blotting on fractions collected from polysome gradients prepared from transgenic testes. Analysis of the endogenous Prm-1 mRNA showed that a fraction of the mRNA sedimented as a 20S particle, most likely representing the translationally repressed form of the mRNA, and that the remainder of the mRNA sedimented in the disome and trisome fraction of the gradient. In contrast, the majority of the transgenic mRNA sedimented as a 20S–40S messenger ribonucleoprotein (mRNP) particle, while only a small fraction sedimented with single ribosomes. Thus, it appears that translational activation of the transgenic mRNA is compromised. Previous analysis of an hGH-Prm-1 3'-UTR transgene that contains just 62 nts of the Prm-1 3'-UTR (Fig. 11.2B) showed that it was properly translationally activated (24). Together, these results suggest that a region of the Prm-1 3'-UTR between nucleotides 38 and 94, is required for wild-type levels of translational activation.

To clone the gene that encodes the 48/50-kDa binding activity, we screened a mouse testis cDNA library using the yeast three-hybrid system. The three-hybrid screen is a modification of the two-hybrid system that allows detection of RNA–protein interaction in yeast (25). As our selectable RNA target we used the first 37 nts of the Prm-1 3'-UTR. Negative control RNAs included a mutant version of the 37-nt RNA, and the 3'-most 23 nts of the Prm-1 3'-UTR, neither of which binds to the 48/50-kDa protein in an RNA EMSA. From this screen we identified one clone that encodes an RNA-binding protein that appears to have specificity for the binding site for the 48/50-kDa protein (Davies and Braun, unpublished data). The RNA-binding protein is a new member of the Y-box family of RNA-binding proteins (26). We have recently shown that an antibody raised against a conserved region of this family of RNA-binding proteins will supershift the 48/50-kDa binding activity in an EMSA (Giorgini and Braun, unpublished data). Therefore, we are confident that the 48/50-kDa complex contains a member of the Y-box family, and it seems likely that the clone we have identified in the three-hybrid screen encodes the family member. At the present time we do not know if the 48- and 50-kDa proteins are encoded by the same gene or by different genes.

Initial studies of the Y box family of RNA-binding proteins suggested that they are components of the translational masking machinery and that they bind RNA nonspecifically. However, Bouvet et al. (27) have shown that at least some members of this family of proteins recognize specific target sequences. We suspect that most if not all these proteins recognize RNA in a sequence-specific manner, but that the target sequences for most of the current family members have not yet been identified.

Two regions of the Prm-1 3'-UTR are capable of repressing translation of a heterologous mRNA in transgenic mice. The first region maps to the first 37 nts of the 3'-UTR and contains the binding site for the 48/50-kDa protein (Fig. 11.2D). The second region is contained within the 3'-most 62 nt of the 3'-UTR (Fig. 11.2B). The 48/50-kDa protein does not bind to the 62-nt region; thus, there must be other translational control factors that mediate the repression through this region. One possiblity is Prbp (protamine RNA-binding protein). We cloned the Prbp gene in an expression screen for cDNAs that encode Prm-1 3'-UTR RNA-binding proteins (28). We screened lambda gt11 cDNA expression libraries prepared with RNA from pachytene spermatocytes, round spermatids, and mixed germ cells with a 156-nt Prm-1 3'-UTR RNA probe labeled with digoxigenin. To enrich for clones that encode Prm-1 RNA-binding proteins, and to minimize the number of clones that encode general RNA-binding proteins, only those plaques that on rescreening did not hybridize with a digoxigenin-labeled 145-nt hGH-3'-UTR RNA probe were retained for subsequent analysis. Of a total of approximately 7×10^5 plaques screened, we identified 19 plaques that bound to the Prm-1 3'-UTR RNA and did not bind, or bound significantly less well, to the hGH 3'-UTR RNA. Characterization of each of the cDNA clones by DNA cross-hybridization revealed that they represented cDNAs from five different genes. The Prbp gene was isolated once from the pachytene spermatocyte library and once from the round spermatid library.

A search of the GenBank database revealed that the Prbp protein has 93% amino acid identity with the human TAR RNA-binding protein (TRBP), strongly suggesting that it is the mouse homologue of TRBP. TRBP is a cellular protein that was cloned first on the basis of its ability to bind to the human immunodeficiency virus type 1 (HIV-1) TAR RNA in vitro (29), and a second time based on its ability to bind to the HIV-1 Rev-responsive element (30). Prbp and TRBP each contain two copies of a lysine- and arginine-rich motif of 24 amino acids that is also found in the human interferon-induced and dsRNA-activated protein kinase, PKR (31), and in numerous other demonstrated or putative dsRNA-binding proteins (32).

To study the in vitro RNA-binding properties of Prbp, a fusion protein between maltose-binding protein (MBP) and Prbp was constructed, expressed in *E. coli,* and affinity purified on amylose resin. In an RNA EMSA, MBP-Prbp binds to the full-length 156-nt Prm-1 3'-UTR RNA producing multiple complexes that all migrate more slowly than the free probe. The MBP-Prbp fusion protein also binds to the 62-nt 3'-most region of the Prm-1 3'-UTR, but does not bind to the 5'-most 82 nt of the Prm-1 3'-UTR RNA. To further map the Prbp-binding site, a series of deletion variants containing different portions of the 3'-most region of the 3'-UTR were constructed and tested in the RNA EMSA. The results of these studies showed that the Prbp protein can bind to two different regions of the 3'-most region of the 3'-UTR. Both of these regions contain putative secondary structure, suggesting that Prbp may recognize structured RNAs. As mentioned, the Prbp protein contains two copies of an amino acid motif found in numerous dsRNA-binding proteins. To determine if MBP-Prbp fusion protein binds to dsRNA, competition assays were performed with poly(I)/poly(C) RNA. Poly(I)/poly(C) RNA competed for binding of radiolabeled Prm-1 3'-UTR RNA, suggesting that Prbp binds dsRNA in vitro.

Western blot analysis of testis protein extracts with anti-MBP-Prbp antibody detected a cluster of four major bands around 40 kDa and a minor band of approximately 33 kDa. No hybridizing bands were detected in testis nuclear extracts. Hybridization of a northwestern (RNA protein) blot containing testis protein extracts with a radiolabled Prm-1 3'-UTR RNA probe revealed the same quartet of bands at 40 kDa, demonstrating that they can all bind to Prm-1 RNA in vitro.

To determine if Prbp is expressed within the somatic or the germ cell compartment of the testis, Western blot analysis was performed on protein extracts prepared from testes of mouse mutants that either lack germ cells entirely (W/W^v) or contain diploid spermatogonial cells but lack meiotic and postmeiotic germ cells because of an arrrest at the beginning of meiosis I (Tfm/Y). None of the four major Prbp proteins of about 40 kDa, or the minor 33-kDa protein, were detected in extracts prepared from either of the mutants, strongly suggesting that expression of Prbp is restricted to the meiotic or postmeiotic germ cells.

To determine the stages and cell type where Prbp is expressed, immunocytochemistry was performed on testis sections using an anti-MBP-Prbp antibody, and serial sections were stained with hematoxylin (a nuclear stain) and periodic acid-Schiff reagent (an acrosomal stain). The Prbp protein was first detected in stage VII pachytene spermatocytes at a very low level. Prbp was also detected in all later stages of primary and secondary spermatocytes and in haploid spermatids through step 10. Immunostaining appeared to be greatest in round spermatids at step 2 to 3. In all cell types that expressed Prbp, the protein appeared to be localized to the cytoplasm, consistent with our earlier Western blot analysis. In summary, the Prbp protein is detected in the cytoplasm of round spermatids through step 10, the cells in which Prm-1 mRNA is repressed, and is not detected in elongated spermatids (steps 11–16), the cells in which Prm-1 mRNA is translated.

The presence of Prbp in the cytoplasmic compartment of round spermatids, coupled with the ability of testis extracts that contain Prbp to bind to exogenously added Prm-1 3'-UTR RNA, suggests that Prbp interacts with Prm-1 mRNA in vivo to inhibit its translation. To test the effect of Prbp on mRNA translation, the MBP-Prbp fusion protein was preincubated with either a chimeric message containing the hGH coding sequence fused to the Prm-1 3'-UTR or an hGH mRNA containing its normal 3'-UTR, and was translated in a wheat germ cell-free translation lysate. Dose–response experiments revealed that both messages, the Prm-1 3'-UTR-containing message and the hGH 3'-UTR-containing message, were sensitive to inhibition in the presence of MBP-Prbp. Translation was not significantly inhibited in the presence of MBP alone or in protein storage buffer alone. These results suggest that Prbp is capable of inhibiting translation in vitro but that it is not specific to mRNAs containing the Prm-1 3'-UTR.

As mentioned earlier, Prbp is striking in its similarity (93% identical at the amino acid level) to human TRBP and is most likely its mouse homologue. The function of TRBP in noninfected cells is unknown, although it has been suggested that TRBP may act as an inhibitor of the interferon-induced protein kinase PKR, and in doing so indirectly controls protein synthesis (30). PKR is a host defense protein whose function in virus-infected cells is to phosphorylate the translation initiation factor eIF-2α (33, 34). Phosphorylation of eIF-2α prevents the obligate exchange of GDP for GTP on eIF-2, resulting in an arrest of protein synthesis and attenuation of virus replication (35). Not unexpectedly, some viruses have evolved mechanisms to elude the host shutdown of protein synthesis by PKR. Among these is vaccinia virus that encodes an inhibitor of PKR (36). The vaccinia virus E3L protein, like TRBP and Prbp, contains several copies of the lysine- and arginine-rich dsRNA-binding motif. E3L is thought to prevent host shutdown of protein synthesis by binding to dsRNA and preventing the activation of PKR. Intriguingly, Park et al. have shown that TRBP can complement a vaccinia virus E3L mutant (30). TRBP has also been shown to interact with PKR in the two-hybrid system and by far-western (protein-protein) blotting, and this interaction is thought to be mediated by binding to a common RNA (37).

Considering the putative role of TRBP as a negative regulator of PKR in human cells, we have entertained a function for Prbp in spermatids in addition to that of a translational repressor of Prm-1 mRNA. It is possible that Prbp functions in spermatids to prevent the activation of PKR. As in a virus-infected cell, new mRNA synthesis in round spermatids is robust, probably partly as a result of the elevation of RNA polymerase II by a factor of 10 fold (1). Perhaps the normal function of Prbp in noninfected cells is to inhibit activation of PKR in cells that are actively differentiating and which are engaged in robust transcription. This model predicts that mutation of Prbp will result in activation of PKR, phosphorylation of eIF-2α, and inhibition of protein synthesis in spermatids. To test this model, mice heterozygous for a null mutation in Prbp have been generated. Analysis of homozygous mutants, providing the mutation does not cause embryonic lethality, allows testing of the various models for the function of Prbp in Prm-1 translational control and for the function of Prbp as a regulator of PKR in spermatids.

Translational Activation

We have determined that there are redundant elements that mediate the translational repression of Prm-1 mRNA. The first of these maps to the 5'-end of the Prm-1 3'-UTR and is bound by proteins of 48/50 kDa (Fig. 11.3). One or both of

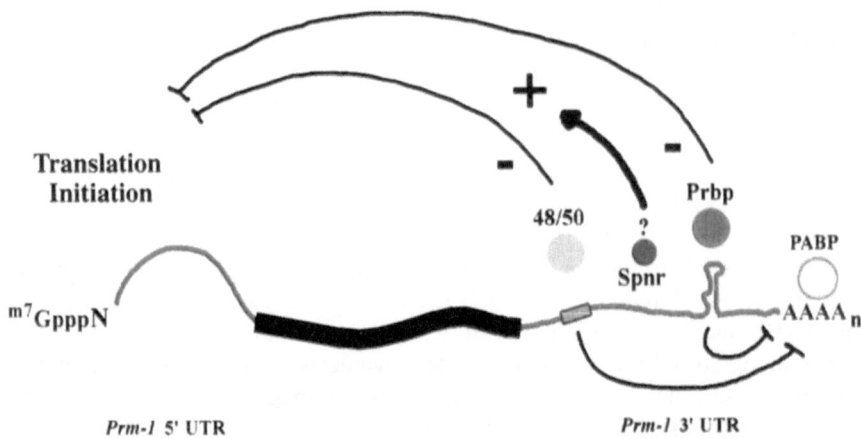

Prm-1 mRNA

FIGURE 11.3. Model shows that the RNA-binding proteins, Prbp (28), and the 48- and 50-kDa RNA-binding complex previously described (23), interact with cis elements in the Prm-1 3'-UTR to repress translation initiation. Possible mechanisms of inhibition include preventing poly(A)-binding protein (PABP) function, or interaction with the 5'-cap to block binding of the 40S ribosomal complex. Translational activation requires sequences present in the 3'-most 62 nts of the Prm-1 3'-UTR and may be dependent on the binding of the microtubule-associated RNA-binding protein, Spnr (9), or an as yet unidentified protein.

these 48/50-kDa proteins is a new member of the Y-box family of RNA-binding proteins. Previous characterization of this family of RNA-binding proteins has suggested that they are nonspecific RNA-binding proteins engaged in translational masking. Our work shows that the 48/50-kDa proteins are sequence-specific RNA-binding proteins, and we suggest that other members of this family may be as well. The 3'-most 62 nts of the Prm-1 3'-UTR is also capable of translational repression of a heterologous mRNA in transgenic mice. This region of the Prm-1 3'-UTR does not contain a binding site for the 48/50-kDa proteins, suggesting that other factors mediate repression through the 3'-end of the 3'-UTR. One candidate protein is Prbp, which binds to a structured region in the 3'-most 62 nts of the Prm-1 3'-UTR, is present at high levels in the cytoplasm of round spermatids and inhibits translation of a heterologous mRNA in vitro. Prbp may also function to inhibit the activation of the dsRNA-activated kinase PKR in hypertranscriptionally active spermatids.

Dissection of the cis-acting sequences required for Prm-1 translational repression in round spermatids led to the discovery that a region of the 3'-UTR is also required for proper translational activation in elongated spermatids. Heterologous reporter mRNAs that contain only the binding site for the 48/50-kDa protein and the poly(A) addition signal are translationally repressed in round spermatids but are poorly activated for translation in elongated spermatids. In contrast, transgenic mRNAs that contain the 3'-most 62 nt of the 3'-UTR are translationally repressed in round spermatids and properly activated for translation in elongated spermatids. We do not know if separate elements in the 62-nt region mediate repression and activation, or if the same element is used in both processes.

How is translational activation mediated? It is not known, but an attractive idea is that activation is coupled to some aspect of spermatid morphogenesis. Possibilities include activation of cytoskeleton-bound mRNPs through extension of the actin-based cytoskeleton in elongated spermatids or recruitment of mRNPs onto the specialized manchette microtubule array in elongating and elongated spermatids. A protein that may be involved in the translational activation of Prm-1 mRNA is Spnr. The gene that encodes the Spnr protein was cloned in the same expression screen in which the Prbp gene was cloned, as described earlier (9). Like Prbp, the Spnr protein contains two copies of the dsRNA-binding motif, and we have shown that Spnr binds to the Prm-1 3'-UTR in vitro. Using confocal immunofluoresence microscopy, we have also shown that Spnr co-localizes to the manchette in elongating and elongated spermatids in wild-type mice and to the abnormal microtubule arrays that are found in several mouse mutants with defective manchettes (Schumacher et al., in manuscript). Spnr can be co-purified with murine testis microtubules, and recombinant Spnr protein binds to testis microtubules in vitro (Schumacher et al., in manuscript). We favor a model in which the Spnr protein binds to Prm-1 mRNPs in elongated spermatids and recruits the message onto polysomes associated with the manchette. Spnr could be interacting with the Prm-1 mRNA through the same structured region in the 3'-most end of the 3'-UTR that Prbp binds, or it could have its own binding site.

Elucidating the mechanism of translational repression in round spermatids and translational activation in elongated spermatids will require identification of the sequences in the mRNA that regulate both processes and of the proteins that bind to them. Testing of models will require genetic experiments in vivo and biochemical experiments in vitro. It is hoped that as we dissect this important form of gene regulation we will also gain greater insights into the complex morphogenesis that makes spermatogenesis one of the most elaborate examples of cellular differentiation.

Acknowledgments. I am indebted to the present and past members of my lab, Keesook Lee, Jill Schumacher, Mark Fajardo, Karen Butner, Jun Zhong, Holly Davies, and Flaviano Giorgini, all of whom have contributed to this work. This work was supported by grants to R.E.B. from the National Institutes of Health (HD27215) and the March of Dimes.

References

1. Schmidt EE, Schibler U. High accumulation of components of the RNA polymerase II transcription machinery in rodent spermatids. Development (Camb) 1995;121:2373–83.

2. Kierszenbaum AL, Tres IL. Structural and transcriptional features of the mouse spermatid genome. J Cell Biol 1975;65:258–70.

3. Monesi V. Ribonucleic acid synthesis during mitosis and meiosis in the mouse testis. J Cell Biol 1964;22:521–32.

4. Bellve AR. The molecular biology of mammalian spermatogenesis. In: Finn CA, ed. Oxford reviews of reproductive biology. Oxford University Press: London, 1979:159–261.

5. Balhorn R, Weston S, Thomas C, Wyrobek AJ. DNA packaging in mouse spermatids. Synthesis of protamine variants and four transition proteins. Exp Cell Res 1984;150: 298–308.

6. Kleene KC, Distel RJ, Hecht NB. Translational regulation and deadenylation of a protamine mRNA during spermiogenesis in the mouse. Dev Biol 1984;105:71–9.

7. Kleene KC, Flynn J. Translation of mouse testis poly(A)+ mRNAs for testis-specific protein, protamine 1, and the precursor for protamine 2. Dev Biol 1987;123:125–35.

8. Mali P, Kaipia A, Kangasniemi M, Toppari J, Sandberg M, Hecht NB, et al. Stage-specific expression of nucleoprotein mRNAs during rat and mouse spermiogenesis. Reprod Fertil Dev 1989;1:369–82.

9. Schumacher JM, Lee K, Edelhoff S, Braun RE. Spnr, a murine RNA-binding protein that is localized to cytoplasmic microtubules. J Cell Biol 1995;129:1023–32.

10. Schumacher JM, Lee K, Edelhoff S, Braun RE. Distribution of Tenr, an RNA binding protein, in a lattice-like network within the spermatid nucleus in the mouse. Biol Reprod 1995;52:1274–83.

11. Braun RE, Peschon JJ, Behringer RR, Brinster RL, Palmiter RD. Protamine 3'-untranslated sequences regulate temporal translational control and subcellular localization of growth hormone in spermatids of transgenic mice. Genes Dev 1989;3:793–802.

12. Lee K, Haugen HS, Clegg CH, Braun RE. Premature translation of protamine 1 mRNA causes precocious nuclear condensation and arrests spermatid differentiation in mice. Proc Natl Acad Sci USA 1995;92:12451–5.

13. Braun RE, Behringer RR, Peschon JJ, Brinster RL, Palmiter RD. Genetically haploid spermatids are phenotypically diploid. Nature (Lond) 1989;337:373–6.

14. Burgos MH, Fawcett DW. Studies on the fine structure of the mammalian testis. I. Differentiation of the spermatids in the cat (Felis domestica). J Biophys Biochem Cytol 1955;1:287–300.

15. Peschon JJ, Behringer RB, Brinster RL, Palmiter RD. Spermatid-specific expression of protamine 1 in transgenic mice. Proc Natl Acad Sci USA 1987;84:5316–9.

16. Zambrowicz BP, Harendza CJ, Zimmermann JW, Brinster RL, Palmiter RD. Analysis of the mouse protamine 1 promoter in transgenic mice. Proc Natl Acad Sci USA 1993;90:5071–5.

17. Meistrich ML. Histone and basic nuclear protein transitions in mammalian spermatogenesis. In: Hnilica G, Stein G, Stein J, eds. Histones and other basic nuclear proteins. Orlando: CRC Press, 1989:165–82.

18. Carré Eusebe D, Lederer F, Le KH, Elsevier SM. Processing of the precursor of protamine P2 in mouse. Peptide mapping and N-terminal sequence analysis of intermediates. Biochem J 1991;277:39–45.

19. Chauviere M, Martinage A, Debarle M, Alimi E, Sautiere P, Chevaillier P. Purification and characterization of precursors of mouse protamine mP2. C R Acad Sci III Life Sci 1991;313:107–12.

20. Morales CR, Kwon YK, Hecht NB. Cytoplasmic localization during storage and

translation of the mRNAs of transition protein 1 and protamine 1, two translationally regulated transcripts of the mammalian testis. J Cell Sci 1991;100:119–31.

21. Tarun SZ Jr, Sachs AB. A common function for mRNA 5' and 3' ends in translation initiation in yeast. Genes Dev 1995;9:2997–3007.

22. Tarun SZ Jr, Sachs AB. Association of the yeast poly(A) tail binding protein with translation initiation factor eIF-4G. EMBO J 1996;15:7168–77.

23. Fajardo MA, Butner KA, Lee K, Braun RE. Germ cell-specific proteins interact with the 3' untranslated regions of Prm-1 and Prm-2 mRNA. Dev Biol 1994;166:643–53.

24. Braun RE. Temporal translational regulation of the protamine 1 gene during mouse spermatogenesis. Enzyme (Basel) 1990;44:120–8.

25. SenGupta DJ, Zhang B, Kraemer B, Pochart P, Fields S, Wickens M. A three-hybrid system to detect RNA-protein interactions in vivo. Proc Natl Acad Sci USA 1996;93: 8496–501.

26. Wolffe AP. Structural and functional properties of the evolutionarily ancient Y-box family of nucleic acid binding proteins. Bioessays 1994;16:245–51.

27. Bouvet P, Matsumoto K, Wolffe AP. Sequence-specific RNA recognition by the *Xenopus* Y-box proteins. An essential role for the cold shock domain. J Biol Chem 1995;270:28297–303.

28. Lee K, Fajardo MA, Braun RE. A testis cytoplasmic RNA-binding protein that has the properties of a translational repressor. Mol Cell Biol 1996;16:3023–34.

29. Gatignol A, Buckler-White A, Berkhout B, Jeang KT. Characterization of a human TAR RNA-binding protein that activates the HIV-1 LTR. Science 1991;251:1597–600.

30. Park H, Davies MV, Langland JO, Chang H, Nam YS, Tartaglia J, et al. TAR RNA-binding protein is an inhibitor of the interferon-induced protein kinase PKR. Proc Natl Acad Sci USA 1994;9:4713–7.

31. Green SR, Mathews MB. Two RNA-binding motifs in the double-stranded RNA-activated protein kinase, DAI. Genes Dev 1992;6:2478–90.

32. St-Johnston D, Brown NH, Gal JG, Jantsch M. A conserved double-stranded RNA-binding domain. Proc Natl Acad Sci USA 1992;89:10979–83.

33. Farrell PJ, Balkow T, Hunt T, Jackson RJ. Phosphorylation of initiation factor eIF-2 and the control of reticulocyte protein synthesis. Cell 1977;11:187–200.

34. Levin D, London IM. Regulation of protein synthesis: activation by double-stranded RNA of a protein kinase that phosphorylates eukaryotic initiation factor 2. Proc Natl Acad Sci USA 1978;75:1121–5.

35. Hershey JW. Translational control in mammalian cells. Annu Rev Biochem 1991;60: 717–55.

36. Chang HW, Watson JC, Jacobs BL. The E3L gene of vaccinia virus encodes an inhibitor of the interferon-induced, double-stranded RNA-dependent protein kinase. Proc Natl Acad Sci USA 1992;89:4825–9.

37. Cosentino GP, Venkatesan S, Serluca FC, Green SR, Mathews MB, Sonenberg N. Double-stranded-RNA-dependent protein kinase and TAR RNA-binding protein form homo- and heterodimers in vivo. Proc Natl Acad Sci USA 1995;92:9445–9.

38. Fajardo MF, Haugen HS, Clegg CH, Braun RE. Separate elements in the 3' untranslated region of the mouse protamine 1 mRNA regulate translational repression and activation during murine spermatogenesis. Dev Biol 1997;191:42–52.

Part IV

Regulation of Cell Death/Aging

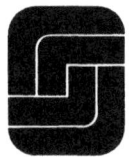

12

Cell Senescence and Aging

Judith Campisi

Replicative Senescence

Most higher eukaryotic cells that can divide in vivo cannot do so indefinitely. Thus, normal cells are said to have a finite division potential or replicative life span. The process that limits the division potential of cells has been termed cellular or replicative senescence. Replicative senescence was first observed when it became possible to culture cells outside the organism. Nearly a century ago, Carrel reported that cell proliferation was unlimited once cells were removed from the organism and cultured ex vivo. This result was sporadically challenged over the ensuing decades. However, it was not until about three decades ago that the finite replicative life span of cells was first formally described. Since then, many cell types from a variety of animal species have been shown to have a finite replicative life span. Most studies of replicative senescence have used cells grown in culture. However, a limited number of studies using cells passaged in intact animals, or cells identified in intact tissues, strongly suggest that cells undergo replicative senescence in vivo (reviewed in 1–3).

Significance of Replicative Senescence

When cells reach the end of their replicative life span, they arrest growth with a G_1 DNA content. This growth arrest is essentially irreversible, and senescent cells cannot be stimulated to enter the S-phase of the cell cycle, much less divide, by any combination of physiological mitogens. The irreversible growth arrest of senescent cells, in conjunction with substantial cell and molecular biological data (2, 4–6), support the idea that replicative senescence is a powerful (but, of course, imperfect) tumor-suppressive mechanism. In addition to arrested growth, recent evidence suggests that senescent cells become resistant to apoptotic death (7). This feature of replicative senescence may at least partially explain the recent finding that senescent cells exist and accumulate in aging human tissue (8). Finally, there is substantial evidence that senescent cells show sometimes striking changes in differentiated function (2, 3, 9–11). The altered function of senescent cells, in conjunction with the evidence that they accumulate in aged tissues and that the number of divisions at which senescence occurs depends inversely on the age of the donor, support the idea that replicative senescence also contributes to organismic aging (3). According to this idea, the accumulation of dysfunctional senescent cells would compromise tissue function and integrity, leading to the decline in tissue homeostasis that occurs with age.

At first glance, the ideas that replicative senescence can contribute to both tumor suppression and aging may seem at odds with each other, or even contradictory. However, evolutionary theories of aging have suggested that some traits that have been selected to optimize health during the period of reproductive fitness can have unselected and deleterious effects later in life. Thus, replicative senescence may have been selected, at least in mammals, to help ensure the relative freedom from cancer that indeed characterizes early and middle adulthood. However, this same process may have deleterious effects later in life because nondividing, dysfunctional senescent cells accumulate. The accumulation of dysfunctional senescent cells may lead not only to the decline in tissue function and integrity that is a hallmark of aging (see 2, 3), but also to the rise in cancer that occurs with age (9). Understanding the causes and effects of cell senescence may be extremely important for understanding whether and how we can prevent the loss of functional and proliferative homeostasis associated with aging.

Exceptions to Replicative Senescence

Do All Cells That Are Capable of Division Senesce?

Because some simple, single-celled organisms, such as *Saccharomyces cerevisiae*, undergo replicative senescence (12), replicative senescence may be a very primitive cellular phenotype. Although there are many parallels between the senescence of unicellular organisms and the replicative senescence of cells from multicellular organisms, there are also some important differences. The most striking difference is that the telomeres, the ends of linear chromosomes, do not shorten during the life span of unicellular organisms. By contrast, telomere shortening appears to invariably occur during the replicative senescence of mammalian cells (at least in human cells, and those rodent species in which telomere shortening can be studied) (11, 12). The focus here is on the causes and consequences of the finite replicative life span of mammalian cells, and, principally, of human cells. Replicative senescence is particularly stringent in human cells. Unlike somatic cells from many rodent species, spontaneous immortalization (that is, spontaneous failure to senesce) is exceedingly rare in human cells (4–6).

Most mammalian cells that are capable of division in vivo (that is, cells that are not terminally differentiated) have been shown to undergo replicative senescence in culture, and are thought to senesce in vivo as well. There are, however, three notable exceptions.

First, the germ line is obviously capable of continuous replication. Although mature sperm and eggs do not proliferate, the cells in the testis and ovary that give rise to mature sperm and ova do not appear to undergo replicative senescence. The division potential of a newly fertilized egg must be equal to that of fertilized eggs from the previous generation, or the species would soon disappear. It follows, then, that somatic cells must acquire a finite replicative life span sometime during embryonic development (see 2, 13, 14). Very little is known about when, where, or how this occurs during development.

Second, some somatic stem cells—most likely in fetal and neonatal tissues, but also possibly in some adult tissues—may be capable of unlimited cell division. This possibility is based primarily on the presence of telomerase activity, which is found mostly but not exclusively in cells that do not appear to undergo replicative senescence (14–19). The idea that the stem cells in selected nonembryonic tissues may have an unlimited replicative life span has not yet been critically tested. Nonetheless, this idea remains an interesting and biologically important prospect, particularly for human cells.

Third, many malignant tumors appear to have cells with an extended or immortal replicative life span. Thus, tumorigenesis frequently selects for cells that have acquired the ability to overcome the proliferative constraints imposed by replicative senescence (2–6). It is now clear that tumorigenesis is a multistep process that entails the accumulation of multiple, successive mutations. Replicative immortality strongly favors the probability that cells will acquire the many mutations that are needed for the development of a full-blown malignant phenotype (4).

The Senescent Phenotype

The Counting Mechanism

The number of divisions that normal cells complete before they senesce depends on the species, age, and genetic background of the donor, as well as the particular cell type (reviewed in 1, 2). This number can be large; for example, 60–80 doublings for fetal or neonatal human cells. There is now substantial, albeit still largely correlative, evidence that the mechanism by which cells sense the number of divisions they have completed depends, at least in part, on the length of their telomeres (6, 11, 20–26).

Telomeres are the ends of linear chromosomes, consisting of the repetitive sequence TTAGGG in humans and other vertebrates, and specialized proteins. The telomeric sequence and its binding proteins form a distinctive structure (27) that prevents chromosome fusions, translocations, and nondysjunctions. Thus, telomeres are essential for maintaining the stability of eukaryotic genomes (28).

Because DNA polymerases are unidirectional and require a labile primer, each round of replication leaves some 3' bases at the telomere unreplicated (21, 24–26). Telomerase, a multimeric enzyme that adds telomeric repeats to chromosome ends de novo (29), is not expressed by most normal somatic cells. Thus, for most cells, the telomeres shorten with each cell cycle. Very little is known, in mammalian cells, about why telomeres should influence cell proliferation or other aspects of the senescent cell phenotype (discussed next) so strongly.

Genetics of Replicative Senescence

The finite replicative life span of cells is a dominant phenotype. This conclusion is based primarily on somatic cell fusion studies in which proliferating normal cells were fused to immortal tumor-derived cells. In the majority of cases, the hybrid cells proliferated for some time, but eventually senesced (1, 2). Moreover, fusions between different immortal human tumor cells often, but not always, produced hybrid cells that senesced (30). These studies strongly suggest that replicative senescence is a genetically dominant trait and that replicative immortality is genetically recessive.

The ability of some, but not all, fusion pairs of immortal human cells to produce hybrids that are *not* immortal has led to the identification of four complementation groups for replicative immortality (31). At present, three of these complementation groups have been assigned to specific human chromosomes. Thus, normal human chromosomes 4, 1, and 7 have been shown to reverse the immortality (that is, induce senescence) of multiple cell lines assigned to the complementation groups B, C, and D, respectively (32–34). The senescence-inducing genes that reside on these chromosomes have not yet been cloned. Taken together, the results of these cell and chromosome fusion studies suggest that at least four distinct, dominant-acting genetic loci act to limit the replicative potential of normal human cells.

The Growth Arrest of Senescent Cells

Upon completing a finite number of divisions, cells arrest growth (used here interchangeably with proliferation) with a G_1 DNA content. Once thus arrested, they cannot be stimulated to proliferate by any known combination of physiological mitogens. This growth arrest, like the finite replicative life span, is a dominant trait in somatic cell hybrids (1, 2). These findings suggest that once cells undergo replicative senescence they express one or more inhibitor of cell-cycle progression that can act in a trans-dominant fashion.

The growth arrest associated with cellular senescence has been studied most extensively in cultured human fibroblasts (1, 2, 9). From these studies, it appears that many genes, including at least three proto-oncogenes, remain mitogen inducible in senescent cells (35, 36). Thus, senescent cells do not fail to proliferate because of a general breakdown of growth factor signal-transducing mechanisms.

Replicative senescence causes the selective repression of a few positive-acting, growth regulatory genes whose expression is important for G_1 progression and DNA synthesis. In fibroblasts, these genes include the c-*fos* proto-oncogene (36), the helix-loop-helix Id-1 and Id-2 genes (37), and the E2F-1 (38) and E2F-5 (39) components of the E2F transcription factor. In addition, the retinoblastoma tumor suppressor protein (pRb) remains in its growth-suppressive form (that is, remains constitutively underphosphorylated) (40). Thus, the immediate cause for the growth arrest of senescent cells appears to be a deficiency in a few, key positive-acting growth regulators.

In addition to deficiencies in positive growth regulators, senescent human fibroblasts overexpress two negative growth regulators: the p21 and p16 inhibitors of cyclin-dependent protein kinases (Cdks) (41, 42). These high levels of p21 and p16 are very likely responsible for the accumulation of inactive Cdk complexes (43), and thus the constitutive underphosphorylation of pRb (40), in senescent cells. Furthermore, because p21 can inhibit E2F by both pRb-dependent and pRb-independent mechanisms (44), the high level of p21 may well be responsible for the lack of E2F activity in senescent cells (38, 39).

p16 and p21 are obvious candidates for dominant inhibitors of cell proliferation expressed by senescent cells. The mechanisms responsible for the senescence-associated increase in p16 and p21 mRNA and protein are not known. In the case of p21, the large increase in mRNA does not appear to result from an increase in transcription (Nakanishi et al., unpublished data). It is unlikely, however, that p16 and p21 are the only growth inhibitors expressed by senescent cells. For example, indirect evidence suggests that senescent cells express at least one additional growth inhibitor, which may belong to a family of basic helix-loop-helix transcription factors (45). Moreover, the upstream regulators of p16 and p21 expression in senescent cells, which have yet to be identified, may induce additional growth inhibitory genes in the cells.

Functional Changes in Senescent Cells

The arrest of cell proliferation is clearly the most important feature of replicative senescence with regard to its role in tumor suppression. However, in addition to an irreversible growth arrest, senescent cells display two other striking phenotypic changes.

First, senescence cells acquire resistance to apoptotic stimuli. The mechanism by which senescent cells resist apoptotic death is not well understood but may involve a stabilization of bcl-2 (7). The fact that senescent cells are in fact resistant to apoptosis underscores the distinction, often blurred in the literature, between cell death and cell senescence.

Second, senescent cells show sometimes striking changes in differentiated functions. The functions that are altered by senescence depend on the cell type. For example, senescent human fibroblasts and endothelial cells overexpress the inflammatory cytokine interleukin-1α (IL-1) (46), and senescent endothelial cells overexpress the cell-specific adhesion molecule I-CAM (47). On the other hand, senescent mammary epithelial cells overexpress the beta isoform of the retinoic acid receptor (48). Very little is known about the mechanisms that alter differentiated gene expression in senescent cells. However, as is the case with growth and differentiation in many mammalian cells, the changes in differentiation appear to be tightly linked to the arrest of cell proliferation.

The changes in cell function that is associated with replicative senescence can have rather profound consequences for cell—and, at least in principle, tissue—function. For example, presenescent dermal fibroblasts express low levels of collagenase and stromelysin, which are metalloproteases (MMPs) that degrade extracellular matrix proteins. They also express relatively high levels of the MMP inhibitors TIMP-1 and TIMP-3 (tissue inhibitor of metalloproteinases 1 and 3). On senescence, MMP expression rises and TIMP expression falls (49–51). Thus, in dermal fibroblasts, replicative senescence entails a fairly dramatic switch in phenotype, from a matrix-producing to a matrix-degrading phenotype.

There are now a number of examples in which senescent cells elaborate cytokines, extracellular matrix-modifying enzymes, or other molecules that can have far-ranging effects on the microenvironment of neighboring cells. In addition to the IL-1 and MMPs discussed here, our preliminary data suggest that senescent fibroblasts overexpress heregulin, an epidermal growth factor- (EGF-) like cytokine that modulates the growth and differentiation of breast and other epithelial cells (Acosta et al., unpublished data). Heregulin is also a potent stimulator of growth for breast epithelial cells that have an amplification of the erbB-2 receptor gene (52). Thus, senescent stromal cells may alter the balance of growth and differentiation of normal epithelium; they may also promote the neoplastic growth of epithelial cells that have acquired potentially oncogenic mutations but have been held in check by their (presenescent) microenvironment (10).

It is not known how the functional changes that accompany replicative senescence occur. Very recently, we have identified a 100-bp transcriptional regulatory element that is responsible for the increase in collagenase expression in senescent human fibroblasts (Testori and Campisi, unpublished data). This element, which we term a senescence-responsive element or SnRE, contains, among other sequences, closely apposed binding sites for the AP-1 and c-ets transcription factors, an arrangement that is found in many cytokine and MMP upstream regions. Whether an SnRE is present and active in the upstream regions of other genes that are overexpressed by senescent cells is not yet known. Nonetheless, our results raise the possibility that the functional changes that occur in senescent cells may be coordinated by one or a few transcriptional regulatory proteins.

Summary

Replicative senescence alters cell growth and cell function. In vivo, replicative senescence may serve to curtail the proliferation of preneoplastic cells. With age, however, senescent cells accumulate in mitotically competent tissues where they may contribute to the decline in tissue function and integrity.

References

1. Stanulis-Praeger B. Cellular senescence revisited: a review. Mech Ageing Dev 1987;38:1–48.
2. Campisi J, Dimri GP, Hara E. Control of replicative senescence. In: Schneider E, Rowe J, eds. Handbook of the biology of aging. New York: Academic Press, 1996:121–49.
3. Campisi J. Replicative senescence: an old lives tale? Cell 1996;84:497–500.
4. Sager R. Senescence as a mode of tumor suppression. Environ Health Perspect 1991;93:59–62.
5. McCormick JJ, Maher VM. Towards an understanding of the malignant transformation of diploid human fibroblasts. Mutat Res 1988;199:273–91.
6. Shay JW, Wright WE. Defining the molecular mechanisms of human cell immortalization. Biochim Biophys Acta 1991;1071:1–7.
7. Wang E. Senescent human fibroblasts resist programmed cell death and failure to suppress bcl2 is involved. Cancer Res 1995;55:2284–92.
8. Dimri GP, Lee X, Basile G, Acosta M, Scott G, Roskelley C, et al. A novel biomarker identifies senescent human cells in culture and aging skin in vivo. Proc Natl Acad Sci USA 1995;92:9363–7.
9. Goldstein S. Replicative senescence: the human fibroblast comes of age. Science 1990;249:1129–33.
10. Campisi J. Aging and cancer: the double-edged sword of replicative senescence. J Am Geriatr Soc 1997;45:482–8.
11. Harley CB, Villeponteau B. Telomeres and telomerase in aging and cancer. Curr Opin Genet Dev 1995;5:249–55.
12. Jazwinski SM. The genetics of aging in the yeast *Saccharomyces cerevisiae*. Genetica (Dordr) 1993;91:35–51.
13. Prowse KR, Greider CW. Developmental and tissue specific regulation of mouse telomerase and telomere length. Proc Natl Acad Sci USA 1995;92:4818–22.
14. Wright WE, Piatyszek MA, Rainey WE, Byrd W, Shay JW. Telomerase activity in human germline and embryonic tissues and cells. Dev Genet 1996;18:173–9.
15. Hiyama K, Hirai Y, Kyoisumi S, Akiyama M, Hiyama E, Piatyszek MA, et al. Telomerase activity in human peripheral blood and bone marrow cells. J Immunol 1995;155:3711–5.
16. Landsdorp PM. Telomere length and proliferation potential of hematopoietic stem cells. J Cell Sci 1995;108:1–6.
17. Broccoli D, Young JW, DeLange T. Telomerase activity in normal and malignant hematopoietic cells. Proc Natl Acad Sci USA 1995;93:9083–6.
18. Buchkovich KJ, Greider CW. Telomerase regulation during entry into the cell cycle in normal human T cells. Mol Biol Cell 1996;7:1443–54.
19. Weng NP, Levine BL, June CH, Hodes RJ. Regulated expression of telomerase activity in human T lymphocyte development and activation. J Exp Med 1996;183:2471–9.
20. Harlebachor C, Boukamp P. Telomerase activity in the regenerative basal layer of the epidermis in human skin and in immortal and carcinoma-derived skin keratinocytes. Proc Natl Acad Sci USA 1996;93:6476–81.
21. Levy MZ, Allsopp RC, Futcher AB, Greider CW, Harley CB. Telomere end-replication problem and cell aging. J Mol Biol 1992;225:951–60.
22. Harley CB, Futcher AB, Greider CW. Telomeres shorten during ageing of human fibroblasts. Nature (Lond) 1990;345:458–60.
23. Hastie ND, Dempster M, Dunlop MG, Thompson AM, Green DK, Allshire RC.

Telomere reduction in human colorectal carcinoma and with ageing. Nature (Lond) 1990;346:866–8.

24. Lindsey J, McGill NI, Lindsey LA, Green DK, Cooke HJ. In vivo loss of telomeric repeats with age in humans. Mutat Res 1991;256:45–8.

25. Allsopp RC, Vaziri H, Patterson C, Goldstein S, Younglai EV, Futcher AB, et al. Telomere length predicts replicative capacity of human fibroblasts. Proc Natl Acad Sci USA 1992;89:10114–8.

26. Allsopp RC, Harley CB. Evidence for a critical telomere length in senescent human fibroblasts. Exp Cell Res 1995;219:130–6.

27. Tommerup H, Dousmanis A, De Lange T. Unusual chromatin in human telomeres. Mol Cell Biol 1994;14:5777–85.

28. Blackburn EH. Structure and function of telomeres. Nature (Lond) 1991;350:569–73.

29. Blackburn EH. Telomerases. Annu Rev Biochem 1992;61:113–29.

30. Pereira-Smith OM, Smith JR. Evidence for the recessive nature of cellular immortality. Science 1983;221:964–7.

31. Pereira-Smith OM, Smith JR. Genetic analysis of indefinite division in human cells: identification of four complementation groups. Proc Natl Acad Sci USA 1988;85: 6042–6.

32. Ning Y, Weber JL, Killary AM, Ledbetter DH, Smith JR, Pereira-Smith OM. Genetic analysis of indefinite division in human cells: evidence for a senescence-related gene(s) on human chromosome 4. Proc Natl Acad Sci USA 1991;88:5635–9.

33. Hensler P, Annab LA, Barrett JC, Pereira-Smith OM. A gene involved in control of human cellular senescence on human chromosome 1q. Mol Cell Biol 1994;14:2292–7.

34. Ogata T, Ayusawa D, Namba M, Takahashi E, Oshimura M, Oishi M. Chromosome 7 suppresses indefinite division of nontumorigenic immortalized human fibroblast cell lines KMST-6 and SUSM-1. Mol Cell Biol 1993;13:6036–43.

35. Rittling SR, Brooks KM, Cristofalo VJ, Baserga R. Expression of cell cycle dependent genes in young and senescent WI38 fibroblasts. Proc Natl Acad Sci USA 1986;83: 3316–20.

36. Seshadri T, Campisi J. c-*fos* repression and an altered genetic program in senescent human fibroblasts. Science 1990;247:205–9.

37. Hara E, Yamaguchi T, Nojima H, Ide T, Campisi J, Okayama H, et al. Id related genes encoding helix loop helix proteins are required for G_1 progression and are repressed in senescent human fibroblasts. J Biol Chem 1994;269:2139–45.

38. Dimri GP, Hara, E, Campisi J. Regulation of two E2F related genes in presenescent and senescent human fibroblasts. J Biol Chem 1994;269:16180–6.

39. Good LF, Dimri GP, Campisi J, Chen KY. Regulation of dihydrofolate reductase gene expression and E2F components in human diploid fibroblasts during growth and senescence. J Cell Physiol 1996;168:580–8.

40. Stein GH, Beeson M, Gordon L. Failure to phosphorylate the retinoblastoma gene product in senescent human fibroblasts. Science 1990;249:666–9.

41. Noda A, Ning Y, Venable SF, Pereira-Smith OM, Smith JR. Cloning of senescent cell derived inhibitors of DNA synthesis using an expression screen. Exp Cell Res 1994;211:90–8.

42. Hara E, Smith R, Parry D, Tahara H, Peters G. Regulation of p16 (CdkN2) expression and its implications for cell immortalization and senescence. Mol Cell Biol 1996;16: 859–67.

43. Dulic V, Drullinger LF, Lees E, Reed SI, Stein GH. Altered regulation of G_1 cyclins in

senescent human diploid fibroblasts: accumulation of inactive cyclin E-cdk and cyclin D-cdk complexes. Proc Natl Acad Sci USA 1993;90:11043–8.

44. Dimri GP, Nakanishi M, Desprez PY, Smith JR, Campisi J. Inhibition of E2F activity by the p21 inhibitor of cyclin-dependent protein kinases in cells expressing or lacking a functional retinoblastoma protein. Mol Cell Biol 1996;16:2987–97.

45. Hara E, Uzman JA, Dimri GP, Nehlin JO, Testori A, Campisi J. The helix-loop-helix protein Id-1 and a retinoblastoma protein binding mutant of SV40 T antigen synergize to reactivate DNA synthesis in senescent human fibroblasts. Dev Genet 1996;18:161– 72.

46. Maier JAM, Voulalas P, Roeder D, Maciag T. Extension of the life-span of human endothelial cells by an interleukin-1a antisense oligomer. Science 1990;249:1570–4.

47. Maier JAM, Statuto M, Ragnoti G. Senescence stimulates U037-endothelial cell interactions. Exp Cell Res 1993;208:270–4.

48. Swisshelm K, Ryan K, Lee X, Tsou H, Peacocke M, Sager R. Down-regulation of retinoic acid receptor β in mammary carcinoma cell lines and its up-regulation in senescing normal mammary epithelial cells. Cell Growth Differ 1994;5:133–41.

49. West MD, Pereira-Smith OM, Smith JR. Replicative senescence of human skin fibroblasts correlates with a loss of regulation and overexpression of collagenase activity. Exp Cell Res 1989;184:138.

50. Millis AJ, Hoyle M, McCue HM, Martini H. Differential expression of metalloproteinase and tissue inhibitor of metalloproteinase genes in aged human fibroblasts. Exp Cell Res 1992;201:373–9.

51. Wick M, Burger C, Brusselbach S, Lucibello FC, Muller R. A novel member of human tissue inhibitor of metalloproteinases (TIMP) gene family is regulated during G_1 progression, mitogenic stimulation, differentiation and senescence. J Biol Chem 1994;269:18953–60.

52. Lupu R, Cardillo M, Cho C, Harris L, Hijazi M, Perez C, et al. The significance of heregulin in breast cancer tumor progression and drug resistance. Breast Cancer Res Treat 1996;38:57–66.

13

Testicular Aging: Leydig Cells and Spermatogenesis

HAOLIN CHEN, LINDI LUO, AND BARRY R. ZIRKIN

Significant age-related decreases in serum testosterone concentration occur in both humans and rats (1). Such changes could result from reduced numbers of Leydig cells, decreased testosterone production by Leydig cells, increased testosterone metabolism, or all of these. Herein we discuss evidence showing that the ability of the Leydig cells of Brown Norway rats to produce testosterone declines with age, and we speculate as to the mechanisms by which this might occur. As in the human (2–8), degenerative changes in the seminiferous tubules accompany aging of the Brown Norway rat (9–11). Age-related atrophy of the seminiferous epithelium (i.e., loss of germ cells) begins focally and subsequently involves most of the testis. We provide evidence that changes in follicle-stimulating hormone (FSH) and intratesticular testosterone concentration are unlikely to cause age-related germ cell loss, and we speculate as to alternative mechanisms.

Leydig Cell Steroidogenesis

Evidence for age-related reductions in Leydig cell testosterone production has come in part from the observation that the capacity of the testes of Brown Norway rats to produce testosterone is significantly reduced in rats of 18 months of age, with further decreases through 30 months (12). These reductions are not the result of loss of Leydig cells; stereological analysis revealed no changes in Leydig cell number between young (6-month-old) and old (18- to 24-month-old) rats (10, 13). Leydig cell volume, however, is reduced in old rats (10, 13), suggesting that aging results in reductions in testosterone production by individual Leydig cells (13). Indeed, Leydig cells, when isolated from the testes of old rats and incubated with increasing concentrations of luteinizing hormone (LH), were found to produce significantly less testosterone than cells isolated from young rats (13, 14).

What are the age-related changes in Leydig cells that explain their reduced ability to produce testosterone? Perturbation of any of the major steps in Leydig cell steroidogenesis (Fig. 13.1), including LH binding to its receptor, cAMP production, cholesterol translocation to the mitochondria, pregnenolone production in the mitochondria, or the subsequent conversion of pregnenolone to progestosterone, 17-hydroxyprogesterone, androstenedione, and testosterone in the smooth endoplasmic reticulum, could result in reduced testosterone production. Testicular LH receptor numbers and affinity (10), and Leydig cell LH-responsive cAMP production (14), have been shown to not decrease with age. In contrast, when young and old Leydig cells were incubated either with LH or with dbcAMP, the old cells produced less testosterone than the young (13). These observations suggest that age-related reductions in testosterone production result from Leydig cell deficits distal to the LH receptor–cAMP cascade.

As yet, the ability of old versus young cells to translocate cholesterol to the mitochondria has not been examined. However, the activities of each of the steroidogenic enzymes, including P-450 side-chain cleavage, 3β-hydroxysteroid dehydrogenase, P-450 17α-hydroxylase/C17–20 lyase, and 17-ketosteroid reductase have been shown to decrease with age (15). Additionally, the ability of Leydig cells to produce NADPH, a cofactor in the steroidogenic pathway, also decreases with age (unpublished data). These observations indicate that aging influences the ability of Leydig cells to convert cholesterol to testosterone at least

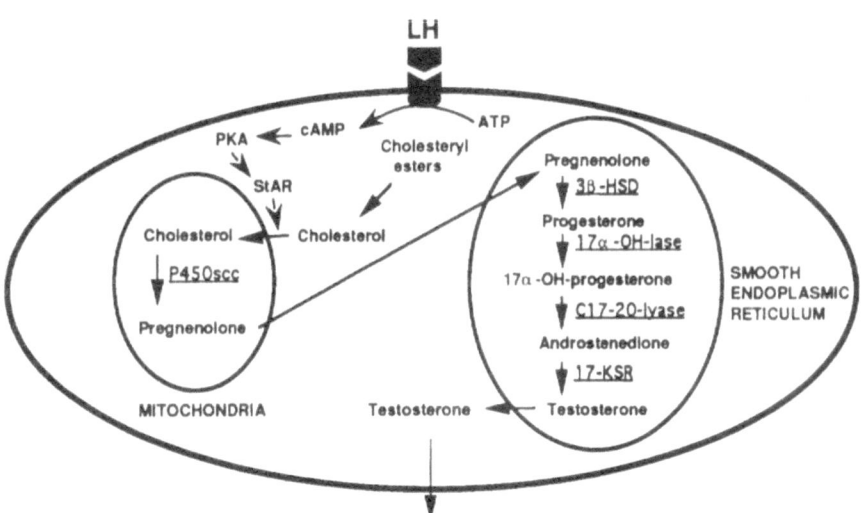

FIGURE 13.1. Diagram of luteinizing hormone (LH) acute regulation of rat Leydig cell steroidogenesis. PKA, protein kinase A; StAR, steroidogenic acute regulatory protein; P-450$_{scc}$, cholesterol side-chain cleavage; 3β-HSD, 3β-hydroxysteroid dehydrogenase; 17-KSR, 17-ketosteroid reductase.

in part because of deficits at multiple points along the steroidogenic pathway, not because of differential regulation of the enzymes that are involved.

What causes aging Leydig cells to become hypofunctional? Many of the changes that Leydig cells undergo as they age, including reduced steroidogenesis, reduced cell volume, and reduced steroidogenic enzyme activities, are similar to the changes that young rat Leydig cells undergo in response to in vivo LH suppression (16). This suggests that changes in LH might be involved in Leydig cell aging. Indeed, the possibility of chronic understimulation by LH is consistent with a number of studies that have reported that serum LH levels decline with age in some rat strains (17–22), and that in some cases age-related reductions in serum testosterone levels can be partially or totally reversed with gonadotropin (23–26). However, in contrast to expectations from these and other studies, we have shown that serum LH levels in young and old Brown Norway rats do not differ (13, 27), and in a recent study, we reported that exogenously administered LH failed to restore testosterone production by aged cells to the level of young cells (Fig. 13.2). Taken together, these results suggest, although do not prove, that LH

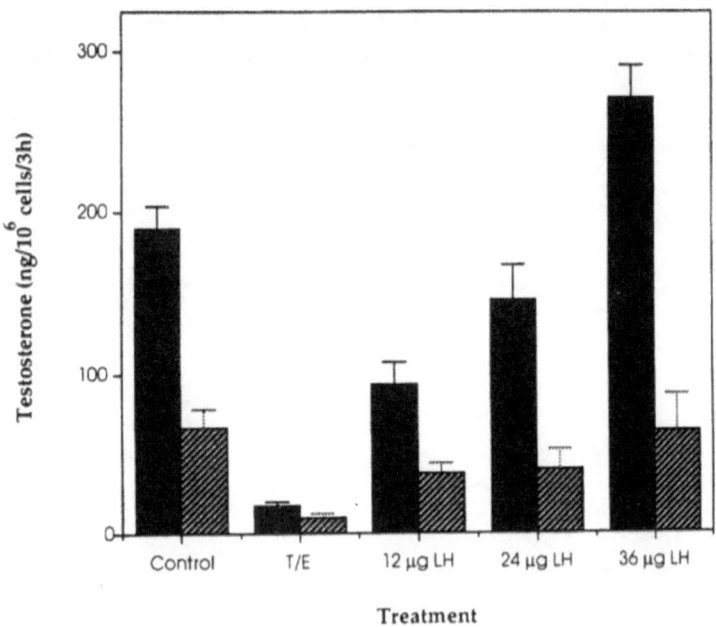

FIGURE 13.2. Testosterone production by Leydig cells isolated from testes of young and aged rats and incubated with maximally stimulating LH (100 ng/ml). Rats received testosterone- and estradiol-containing Silastic implants to suppress endogenous LH and, at the same time, miniosmotic pumps designed to deliver 0–36 μg LH/day in pulses for 5 days. Leydig cells were isolated from 10–12 testes from each treatment group. Data represent the mean ± SEM for three cell preparations per treatment group.

changes if any may not be responsible for the reduced ability of aged Leydig cells to produce testosterone.

Whether or not LH is involved, the possibility remains that Leydig cells become steroiodogenically hypofunctional as a consequence of age-related changes extrinsic to the Leydig cells themselves. To address this possibility, we asked whether young Leydig cells would continue to produce high levels of testosterone if they were situated in aged testes. We reasoned that if they did so, the explanation would be that Leydig cell aging results from factors intrinsic to Leydig cells themselves. To address this, young and old rats were injected with ethane dimethanesulfonate (EDS), an alkylating agent that had been shown to eliminate Leydig cells from young testes (28, 29). We found that EDS also eliminated Leydig cells from aged testes (by 1 week), and that subsequently (by 5–10 weeks) new Leydig cells repopulated the testes at both ages (27).

At 10 weeks, when Leydig cell number in young and old testes was equivalent (70% of control), the testes at both ages produced equivalent amounts of testosterone, at the high level of young control testes. This level significantly exceeded that produced by old control testes (Fig. 13.3). Consistent with this, the ability of equal numbers of Leydig cells isolated from the testes of young and old EDS-treated rats to produce testosterone also was equivalent at 10 weeks, with levels per cell even higher than the level produced by young control cells. Thus, although situated in the aged testes of aged rats, and despite the environment created by an aged hypothalamic-pituitary axis, the steroidogenic function of Leydig cells

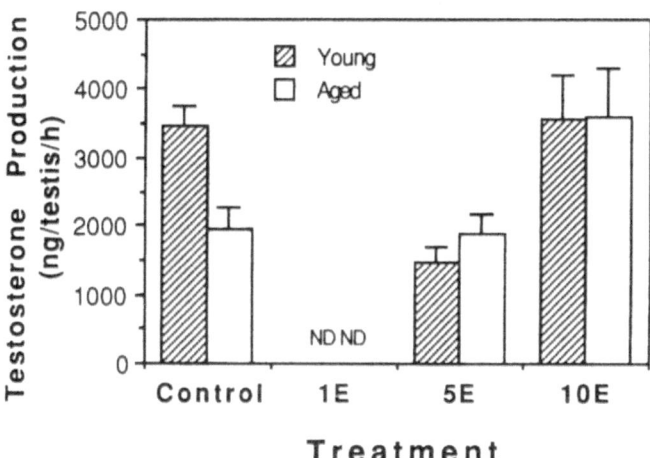

FIGURE 13.3. Testosterone production by testes from rats of 3–6 (young) (hatched bars) and 18–21 (aged) (white bars) months of age perfused in vitro with maximally stimulating LH. Rats received a single injection of ethane dimethanesulfonate (EDS) and were sacrificed 1 (1E), 5 (5E), or 10 (10E) weeks thereafter. By 10 weeks, the testes of young and aged EDS-treated rats produced testosterone at the high level of young controls, a level that significantly exceeded that of aged controls. From Chen et al. (27), with permission.

restored to old rat testes was equivalent to that of Leydig cells restored to young testes. How long the new cells in the old testes retain their steroidogenic "youth" is not known, and this obviously is a critical issue that must be examined.

At this juncture, it is fair to say that we know a great deal about the differences between young and old Leydig cells but little about the mechanisms by which aging Leydig cells become hypofunctional. Despite the evidence just outlined, age-related changes in LH cannot yet be ruled out. For example, the form of the LH signal itself (e.g., its pulsatility), and the ability of LH to enter the interstitial compartment of the testis and to reach and thus stimulate Leydig cells, may change with age and thereby affect steroidogenesis. Another possible mechanism to explain decreased steroidogenic function is oxidative damage to Leydig cell proteins and lipids. In other systems, oxidative damage increases with age in response to the continuous oxidative stress to which aerobic systems are subjected (30). There is evidence from in vitro studies that reactive oxygen species can inhibit Leydig cell steroid production through adverse effects on steroidogenic enzymes (30–32). Moreover, there is evidence that such a mechanism may be physiologically relevant for Leydig cells because oxygen free radicals are produced during the steroidogenic steps regulated by P-450 enzymes (33). To this point, however, the possibility that oxidative damage is involved in Leydig cell aging has received little serious attention.

Age-Related Changes in Spermatogenesis

During aging of both humans and rats, focal atrophy of the seminiferous epithelium is seen (2, 3, 34). In the Brown Norway rat, there is an initial loss of germ cells during the period from 18 to 24 months of age (9–11), and this continues over time, involving more and more of the seminiferous epithelium and thereby causing progessive losses in testis weight (Fig. 13.4). By 18–24 months, some of the testes are entirely normal, while others are partially regressed, with atrophic and normal seminiferous tubules coexisting therein (Fig. 13.5). By 30 months, all testes are almost entirely atrophic, with most tubules containing Sertoli but few or no germ cells. Loss of Sertoli cells also has been reported in old rats (10).

Do age-related changes in spermatogenesis result from hormonal changes? The two major hormones that regulate spermatogenesis are FSH and testosterone. In Brown Norway rats, as in the human, serum FSH levels increase significantly with age (10, 11). The fact that FSH rises as progressive losses of germ cells are occurring makes it unlikely that changes in FSH are responsible for germ cell loss.

Are changes in testosterone responsible for germ cell loss? Both serum and intratesticular testosterone concentrations decrease with age (12). Are decreases in intratesticular testosterone responsible for the age-related losses of germ cells? Figure 13.4 shows that in rats of 23 months, about half the testes are at least partially regressed. We know from histological examination (Fig. 13.5) that this occurs primarily because of germ cell loss. Figure 13.6 compares the testosterone concentration in the seminiferous tubule fluid of testes from young (6-month-old)

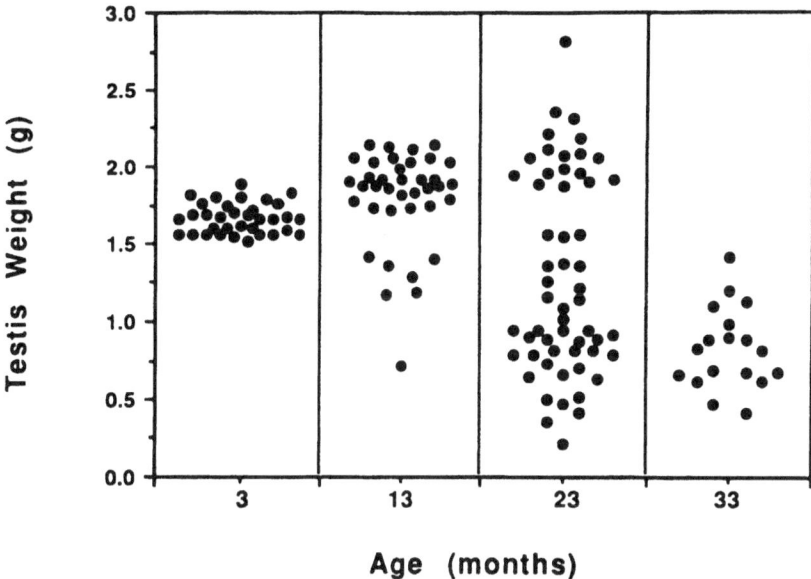

FIGURE 13.4. Individual testis weights of rats of 3, 13, 23, and 33 months of age. Note that by 23 months about half the testes weighed less than 1.5 g and half weighed 1.5 g or more.

and old (20-month-old) rats; the normal (≥ 1.5 g; big) and partially or fully regressed (< 1.5 g; small) testes of the 20-month-old rats were assessed separately. The equivalence of testosterone concentration in the germ cell-containing (big) and germ cell-deficient (small) testes of the old rats suggests strongly that diminished intratesticular testosterone cannot explain the age-related loss of germ cells. We concluded that it is unlikely that changes either in FSH or in testosterone cause age-related losses of germ cells.

Are nonendocrine factors involved in age-related germ cell losses? A variety of experimentally induced pathological conditions can effect germ cell loss and thus result in tubular atrophy. These include ischemia, increase in testicular temperature, and autoimmune disease. Whether or not age-related failure of spermatogenesis results from pathology such as this is unknown, although there is some evidence that argues against this concept. First, germ cell losses typically begin in one testis, without left–right bias. The vasculature (venous drainage) differs between the two testes (35), and thus left–right bias, if it occurred, would suggest the possibility of specific vascular disease. Indeed, we have observed no significant vascular disease in any organ of aged Brown Norway rats, including the testis (12). Also, seminiferous tubule atrophy does not appear to result from an acute autoimmune disease; we have looked for, but have not observed, accumulation of lymphocytes around tubules undergoing atrophy.

An alternative explanation for age-related losses of germ cells is changes in factors intrinsic to the germ cells themselves or to the Sertoli cells with which they

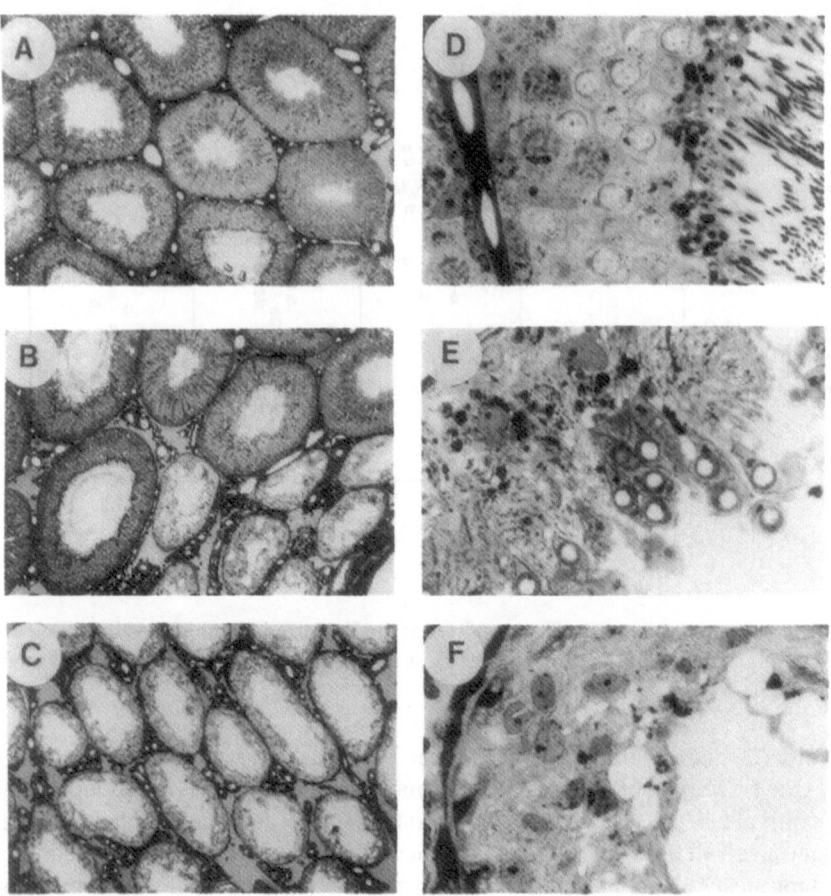

FIGURE 13.5. The effects of aging on the morphology of the seminiferous epithelium. (*A,D*) Seminiferous tubules, young rat. (*B,E*) Adjacent normal and atrophic tubules, old rat. (*C,F*) Fully atrophic tubules, old rat. From Wright et al. (9), with permission.

associate. For example, programmed genetic changes might result in the apoptotic death of germ cells, including stem spermatogonia. Another possibility is that the ability of the germ cells to resist or repair damage from toxic agents might be compromised. For example, free radical damage might occur over time, irreversibly damaging germ cells and ultimately causing their death. These issues are currently under investigation in our laboratory.

Summary

Serum testosterone concentration and Leydig cell steroidogenesis decrease with age in Brown Norway rats. Age-related changes in Leydig cell function involve

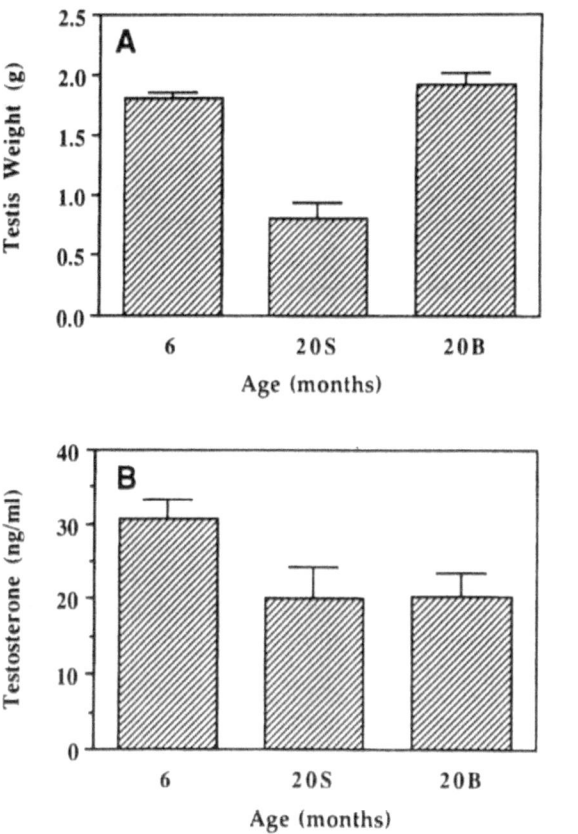

FIGURE 13.6. Testis weight (A) and testosterone concentration in the seminiferous tubule fluid (B) in rats of 6 ($n = 22$) and 20 ($n = 14$) months of age. S, small testes from 20-month-old rats, weighing less than 1.5 g ($n = 6$); B, big (normal) testes from 20-month-old rats, weighing 1.5 g or more ($n = 8$).

reductions in all the steroidogenic enzymes responsible for converting cholesterol to testosterone. At this point, we do not know whether factors intrinsic or extrinsic to the Leydig cells are involved in their aging. Spermatogenesis also is compromised with age, with initial focal and later extensive losses of germ cells characterizing the aging testis. Germ cell losses appear to be unrelated to changes either in FSH or testosterone. Studies currently under way seek to understand the mechanisms by which Leydig cells age, and the effect, if any, of Leydig cell aging on spermatogenesis.

Acknowledgment. This work was supported by NIH grant AG08321.

References

1. Chen H, Luo L, Zirkin BR. Leydig cell structure and function during aging. In: Payne A, Hardy MP, Russell LD, eds. The Leydig cell. Vienna, IL: Cache River Press, 1996:221–30.
2. vom Saal FS, Finch CE. Reproductive senescence; phenomena and mechanisms in mammals and selected vertebrates. In: Knobil E, Neill J, eds. The physiology of reproduction, Vol. 2. New York: Raven Press, 1988:2351–413.
3. Paniagua R, Nistal M, Saez FJ, Fraile B. Ultrastructure of the aging human testis. J Electron Microsc Tech 1991;19:241–60.
4. Johnson L, Petty CS, Neaves WB. Further quantification of human spermatogenesis; germ cell loss during postprophase of meiosis and its relationship to daily sperm production. Biol Reprod 1983;29:207–15.
5. Johnson L, Zane RS, Petty CS, Neaves WB. Quantification of human Sertoli cell population: its distribution, relation to germ cell numbers, and age-related decline. Biol Reprod 1984;31:785–95.
6. Johnson L, Brumbles JS, Bagheri A, Petty CS. Increased germ cell degeneration during postprophase of meiosis is related to increased serum follicle-stimulating hormone concentrations and reduced daily sperm production in aged men. Biol Reprod 1990;42:281–7.
7. Paniagua R, Martin A, Nistal M, Amat P. Testicular involution in elderly men: comparison of histologic quantitative studies with hormone patterns. Fertil Steril 1987;47: 671–9.
8. Paniagua R, Nistal M, Amat P, Rodriguez MC, Martin A. Seminiferous tubule involution in elderly men. Biol Reprod 1987;36:939–47.
9. Wright WW, Fiore C, Zirkin BR. The effect of aging on the seminiferous epithelium of the Brown Norway rat. J Androl 1993;14:110–7.
10. Wang C, Leung A, Sinha-Hikim AP. Reproductive aging in the male Brown-Norway rat: a model for the human. Endocrinology 1993;133:2773–81.
11. Gruenewald DA, Naai MA, Hess DL, Matsumoto AM. The Brown Norway rat as a model of male reproductive aging: evidence for both primary and secondary testicular failure. J Gerontol 1994;49:B42–B50.
12. Zirkin BR, Santulli R, Strandberg JD, Wright WW, Ewing LL. Testicular steroidogenesis in the aging Brown Norway rat. J Androl 1993;14:118–23.
13. Chen H, Hardy MP, Huhtaniemi I, Zirkin BR. Age-related decreased Leydig cell testosterone production in the Brown Norway rat. J Androl 1994;15:551–7.
14. Zirkin BR, Chen H, Luo L. Leydig cell steroidogenesis in aging rats. Exp Gerontol 1997;32:529–37.
15. Luo L, Chen H, Zirkin BR. Are Leydig cell steroidogenic enzymes differentially regulated during aging? J Androl 1996;17:509–15.
16. Ewing LL, Zirkin BR. Leydig cell structure and steroidogenesis function. Recent Prog Horm Res 1983;39:599–635.
17. Karpas AE, Bremner WJ, Clifton DK, Steiner RA, Dorsa DM. Diminished luteinizing hormone pulse frequency and amplitude with aging in the male rat. Endocrinology 1983;112:788–92.
18. Steiner RA, Bremner WJ, Clifton DK, Dorsa DM. Reduced pulsatile luteinizing hormone and testosterone secretion with aging in the male rat. Biol Reprod 1984;31: 251–8.
19. Shaar CJ, Euker JS, Riegle GD, Meites J. Effects of castration and gonadal steroids on

serum luteinizing hormone and prolactin in old and young rats. J Endocrinol 1975;66: 45–51.

20. Gray GD. Changes in the levels of luteinizing hormone and testosterone in the circulation of ageing male rats. J Endocrinol 1978;76:551–2.

21. Bruni JF, Huang HH, Marshall S, Meites J. Effects of single and multiple injections of synthetic GnRH on serum LH, FSH and testosterone in yound and old rats. Biol Reprod 1977;17:309–12.

22. Pirke KM, Geiss M, Sintermann R. A quantitative study on feedback control of LH by testosterone in young adult and old male rats. Acta Endocrinol 1978;89:789–95.

23. Harman SM, Danner RL, Roth GS. Testosterone secretion in the rat in response to chorionic gonadotropin: alterations with age. Endocrinology 1978;102:540–4.

24. Miller AE, Riegle GD. Serum testosterone and testicular response to hCG in young and aged male rats. J Gerontol 1978;33:197–203.

25. Pirke KM, Bofilias I, Sintermann R, Langhammer H, Wolf I, Pabst HW. Relative capillary blood flow and Leydig cell function in old rats. Endocrinology 1979;105: 842–5.

26. Tsitouras PD, Kowatch MA, Blackman MR, Harman SM. In vivo chorionic gonadotropin administration reverses the testosterone secretory defect of Leydig cells from old rat. J Gerontol 1984;39:257–63.

27. Chen H, Huhtaniemi I, Zirkin BR. Depletion and repopulation of Leydig cell in the testes of aging Brown Norway rats. Endocrinology 1996;137:3447–52.

28. Kerr JB, Donachie K, Rommerts FFG. Selective destruction and regeneration of rat Leydig cells in vivo. Cell Tissue Res 1985;242:145–56.

29. Bartlett JMS, Kerr JB, Sharpe RM. The effect of selective destruction and regeneration of rat Leydig cells on the intratesticular distribution of testosterone and morphology of the seminiferous epithelium. J Androl 1986;7:240–53.

30. Tabatabaie T, Floyd RA. Protein damage and oxidative stress. In: Holbrook NJ, Martin GR, Lockshin RA, eds. Cellular aging and cell death. New York: Wiley-Liss, 1996:35–49.

31. Quinn PG, Payne AH. Oxygen-mediated damage of microsomal cytochrome P-450 enzymes in cultured Leydig cells. J Biol Chem 1984;259:4130–5.

32. Georgiou M, Perkins LM, Payne AH. Steroid synthesis-dependent, oxygen-mediated damage of mitochondrial and microsomal cytochrome P-450 enzymes in rat Leydig cell cultures. Endocrinology 1987;121:1390–9.

33. Peltola V, Huhtaniemi I, Metsa-Ketela T, Ahotupa M. Induction of lipid peroxidation during steroidogenesis in the rat testis. Endocrinology 1996;137:105–12.

34. Humphreys PN. The histology of the testis in aging and senile rats. Exp Gerontol 1977;12:27–34.

35. Goldstein M, Phillips DM, Sundaram K, Young GPH, Gunsalus GL, Thau R, et al. Surgical transplantation of testes in isogenic rats: method and function. Biol Reprod 1983;28:971–82.

14

Molecular Regulation of Testicular Cell Death

C. Michael Knudson, Kenneth S.K. Tung, and Stanley J. Korsmeyer

Cell death has been known to be a frequent fate of male germ cells. In fact, one estimate suggests that death may be the most common fate of spermatogonia in the mature testis (1). Despite this realization, the molecular regulators and significance of this death are poorly understood. In this chapter, we first review genes that regulate cell death in general and then focus on those genes which participate in male germ cell death.

Molecular Regulators of Cell Death

Cell death can assume the morphological characteristics of either apoptosis or necrosis (2). Necrotic death is typified by rupture of plasma membranes with nuclear and mitochondrial swelling and is often associated with an inflammatory response. In contrast, apoptotic death is characterized by nuclear condensation and chromatin margination, membrane blebbing, and relative preservation of organelles (2). Programmed cell death (PCD) describes cell death that generally requires active participation of the cell in its own death. Cells undergoing PCD have morphological features of apoptosis. The realization that apoptotic death is often "programmed" led to an extensive search for genes that regulate and participate in apoptosis. The nematode *Caenorhabditis elegans* has been especially informative in delineating genes in the apoptotic pathway because the fate of all its somatic cells has been determined and 131 of the 1090 cells undergo apoptotic cell death (3). Mutagenesis of the worm identified three genes that are critical regulators or mediators of all developmental deaths (Fig. 14.1). Loss-of-function mutations in *ced-3* or *ced-4* prevent cell death in all 131 cells (4). In contrast, a gain-of-function mutation in *ced-9* prevents all developmental cell deaths while loss-of-function mutations are maternal effect lethal as the result of ectopic cell death (5, 6). Of these, *ced-3* and *ced-9* are homologous to mammalian genes.

A) C. Elegans

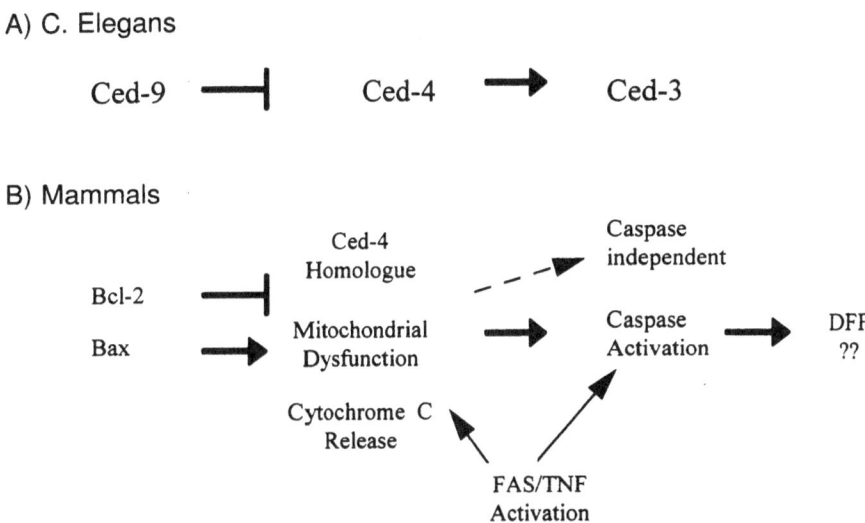

B) Mammals

FIGURE 14.1. Models of apoptosis regulation show, for comparison, the apoptotic pathways for *C. elegans* and mammals. Although similarities exist, mammals may have caspase-independent pathways of apoptosis.

Caspase Family

Cloning of the *ced-3* gene was particularly enlightening as it is homologous to interleukin-1β-converting enzyme (ICE/caspase-1), which had not previously been implicated in regulating or participating in cell death (7). Activated ICE is a heterodimeric cysteine protease required for IL-1β processing and is implicated in other cytokine processing (8). ICE is now known to be part of a family of at least 10 cysteine proteases called caspases because of their unusual specificity for cleavage after aspartic acid (9). ICE itself is unlikely to frequently participate in cell death as ICE-deficient mice do not demonstrate obvious defects in developmental cell death (10, 11). However, multiple lines of evidence implicate caspase activation in mammalian cell death. First, caspase-3 and caspase-7 have been shown to be activated in cells after apoptotic stimuli (12, 13). Furthermore, caspase-3 deficient mice die in utero and showed marked neuronal cell hyperplasia, likely as the result of decreased cell death (14). Viruses contain antiapoptotic genes that are known caspase inhibitors (15–17). Finally, caspase inhibitors block cell death after stimulation of the Fas or TNF receptor (13, 18–21). These results support caspases as important mediators of mammalian apoptosis. Although caspases are essential for cell death in *C. elegans* and are clearly activated in response to many apoptotic signals in mammalian cells, two recent reports have suggested that inhibition of caspases may be insufficient to prevent cell death in response to selected apoptotic stimuli (19, 22). Thus, distinct from the cells of *C.*

elegans, mammalian cells may have caspase-independent pathways of cell death that mediate death downstream from the *Bcl-2* family (see Fig. 14.1).

A critical question is what regulates the activation of caspases. Activated caspases are formed from an inactive precursor protein by cleavage at their own aspartic acid motifs (23). This specificity provides the opportunity for caspase autoactivation and an amplifying cascade (18, 24). However, what initiates the first cleavage of zymogen remains to be firmly established. Signaling through the Fas receptor has been shown to directly recruit caspase-8 to the receptor complex (Fig. 14.1) (25). Thus, recruitment of caspases by protein–protein interaction could also contribute to their activation. *Ced-4*, a requisite gene in *C. elegans* cell death, is a candidate for a caspase activator because it can bind ced-3, caspase-1, and caspase-8 (26), although these protein–protein interactions have not yet been demonstrated to activate the enzymes. Other reports indicate that cytochrome C can activate the caspases in conjunction with other factors (27). Cytochrome C appears to be released from mitochondria in certain apoptotic systems (28, 29). Caspase activation may involve more than one mechanism, which may vary depending on the signal that initiates apoptosis.

An unresolved issue is which of the substrate(s) of the caspases actually mediate the features of apoptosis. Although a large number of substrates have been identified, including PARP, D4-GDI, lamin A, SREBP-1, and SREBP-2, GAS2, and protein kinase Cδ, the role of each of these in apoptosis in not clear (30). A recent substrate for caspase-3, DFF (31), is implicated in mediating DNA fragmentation in apoptotic cells. Caspase inhibitors were unable to block DNA fragmentation of nuclei treated with DFF activated by caspase preincubation (Fig. 14.1). Whether DFF is capable of mediating other features of apoptosis such as nuclear condensation or membrane blebbing is not known.

BCL-2 Family

The homology of *ced-9* to the *Bcl-2* family demonstrates that apoptosis is a highly conserved evolutionary pathway (32). Substantiating this, *Bcl-2* has been shown to inhibit apoptosis when expressed in *C. elegans* (32, 33). BCL-2 prevents the activation of caspases and thus appears to regulate death upstream of caspases (13, 20, 34). Unlike *C. elegans* which has only a single known *Bcl-2* family member, mammals possess multiple *Bcl-2* family members that can have opposite effects on cell death (35). BCL-2, BCL-X_L, BCL-W, MCL1, A1, and NR-13 have anti-apoptotic activity while BAX, BAK, BCL-X_S, BAD, BIK, and BID have pro-apoptotic activity (35, 36). Of interest, the proapoptotic family member BAX has a propensity to bind to and heterodimerize with the antiapoptotic family members (37). Multiple lines of evidence suggest these proteins act as a rheostat (38) to determine the response of the cell to an apoptotic stimulus. Preset levels of proteins dictate susceptibility, while rapid expression of some pro-apoptotic family members is sufficient to induce apoptosis (19, 39).

The biochemical activity of BCL-2 family members that accounts for their regulation of apoptosis remains an active area of research. Several lines of investigation provide possible roles for BCL-2 activity (see Fig. 14.1). For the worm, *ced-9* binds to *ced-4*, and this protein interaction may account for its activity (26, 40, 41). This is consistent with data indicating *ced-9* protects against apoptosis poorly in a *ced-4*-null background (42). Also, the localization of both cytochrome C and BCL-2 to mitochondria raised the possibility that *Bcl-2* prevents the release of cytochrome C. Other data indicate that the presence of BCL-2 can prevent the mitochondrial dysfunction that follows an apoptotic stimulus (43, 44). Recent structural studies of BCL-X_L revealed similarity to diphtheria toxin and colicins, which have pore-forming capability (45). Consistent with this, purified BCL-X is able to form a cation channel when reconstituted into lipid bilayers (46). How this pore-forming capacity might regulate apoptosis remains to be elucidated.

Cell Death Within the Testis

Cell death within the testis has been recognized since the turn of the century. However, only recently has the extent and type of death been defined. For instance, Huckins used radiolabeling methods to estimate that only 25% of the potential number of preleptotene spermatocytes are derived from spermatogonia in the adult rodent testis (1). He proposed that the ratio of spermatogonia to Sertoli cells may be a limiting factor in germ cell survival. Morphological characterization of spermatogonia supports apoptosis as the major mechanism of death (47). In addition, significant death occurs in more mature spermatocytes and spermatids (47). However, this cell death may be distinct because the cells show changes more characteristic of necrosis (47). Thus, it has been well established that cell death is a common fate of male germ cells and that some of these die by apoptosis. Only recently has the molecular regulation of germ cell death begun to be characterized. Both the Fas receptor and its ligand are necessary for the immune privileged status of the seminiferous tubule (48). The Fas ligand on Sertoli cells is proposed to prevent the migration of Fas receptor expressing lymphocytes into the tubules (48). However, Fas receptor- and Fas ligand-deficient mice are fertile without any overt defects in germ cell apoptosis described (49, 50). Thus, Fas is unlikely to be a normal determinant of developmental germ cell death. In contrast, *Bcl-2* family members have recently been shown to be critical regulators of germ cell death.

Disruption of *Bax*, a proapoptotic member of the BCL-2 family (51), results in male infertility. *Bax*-deficient mature mice demonstrate paradoxical testicular hypoplasia (Fig. 14.2B) and increased cell death by TUNEL (52). The testis is characterized by many tubules with an accumulation of monotonous spermatocytes extending several cell layers out from the basal lamina. These cells are germ cell nuclear antigen 1 (GCNA1) positive but LDHC4 negative, consistent with their identification as immature germ cells (52). The paradoxical increased death could have multiple explanations. First, *Bax* may occasionally display antiapoptotic activity in a cell-type specific manner. This possibility is not unprecedented, as several genes that regulate cell death including *Bcl-x, ced-4, ced-9,* and *caspase-2*

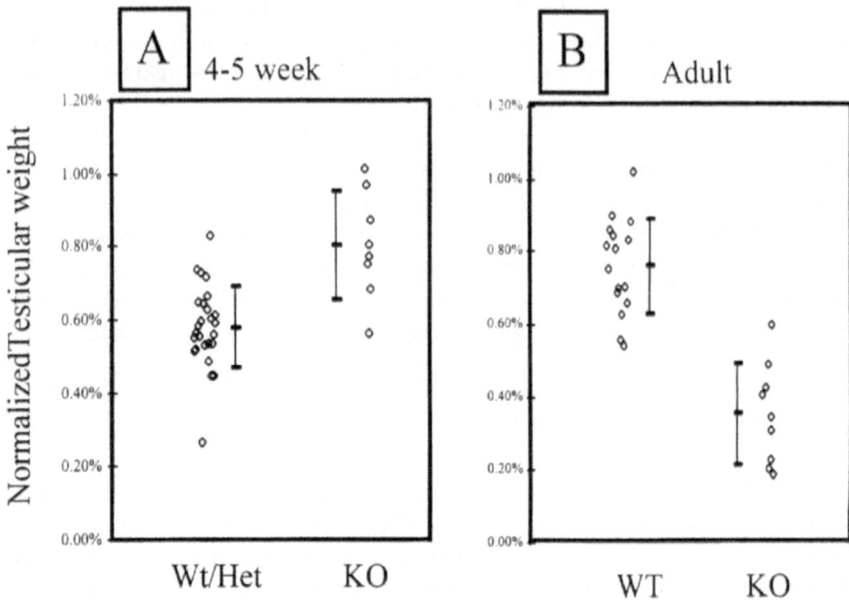

FIGURE 14.2. *Bax* deficient mice develop testicular hyperplasia before testicular atrophy. Normalized testicular weights from 4–5-week-old mice (*A*) or adult mice more than 5 weeks of age (*B*) are from control (WT or Het) and *Bax*-deficient (KO) mice. Testicular weights were normalized by dividing the weight of both testes by total body weight. Individual mice are shown by diamonds with the mean ± SD adjacent to the raw data. In both groups, the *Bax*-deficient mice are significantly different from the control group (*P* < .05, Student's *t*-test).

all have been proposed to have both anti- and proapoptotic activity (53–55). Alternatively, *Bax* might be required for germ cell maturation, and increased death in the adult would result from the disruption of maturation.

To further address this, *Bax*-deficient mice were examined before adulthood. *Bax*-deficient pubertal mice (4–5 weeks) demonstrate testicular hypertrophy (Fig. 14.2A) and also have markedly hypercellular seminiferous tubules (unpublished observations). Furthermore, analysis of cell death using TUNEL demonstrates decreased spermatocyte apoptosis at 1–1.5 weeks of age (unpublished observations). These observations are consistent with a *Bax*-dependent wave of germ cell death, which is essential for normal testicular development. To examine whether the testicular defect in *Bax*-deficient mice is caused by the unopposed activity of BCL-2, double-deficient mice were generated and spermatogenesis was examined. Although *Bcl-2*-deficient mice are severely runted with defects in renal function and lymphocyte survival (56), they are fertile and have normal spermatogenesis (Fig. 14.3). However, *Bcl-2/Bax* double-deficient mice have severely disrupted spermatogenesis that appears identical to the *Bax*-deficient testis (Fig.

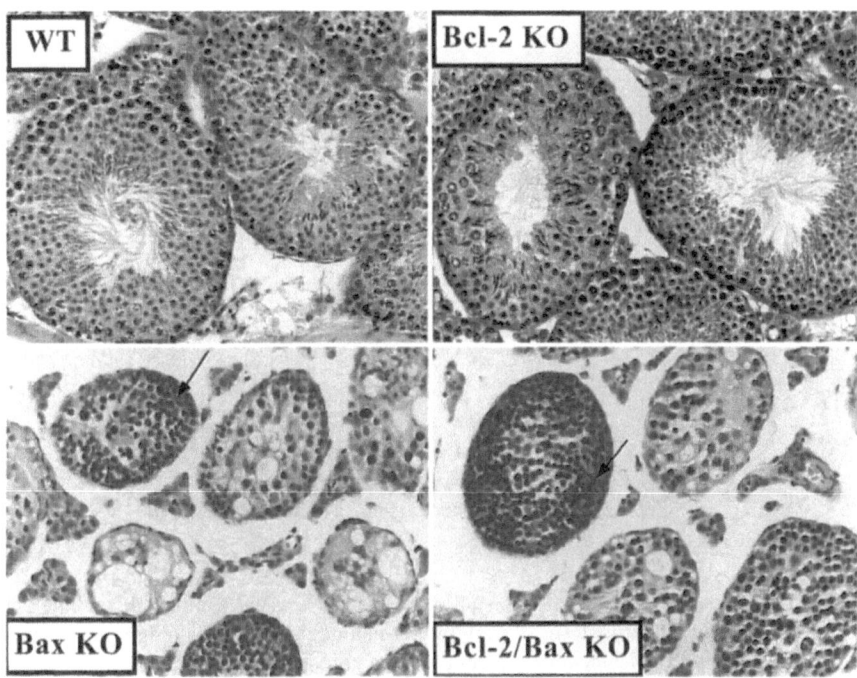

FIGURE 14.3. *Bax*-deficient infertility is not caused by unopposed *Bcl-2*. Testes from control (WT), *Bcl-2*-deficient (*Bcl-2* KO), *Bax*-deficient (*Bax* KO), and *Bcl-2/Bax* double-deficient mice (*Bcl-2/Bax* KO) show that the abnormal germ cell accumulation seen in the *Bax* KO mice is clearly evident in double-deficient mice (arrows). Hematoxylon-eosin (H–E) staining.

14.3) These results demonstrate that *Bax*-deficient infertility is not caused by the activity of *Bcl-2*.

In contrast to the *Bcl-2* deficient mice, transgenic mice that overexpress BCL-2 within germ cells are infertile with testicular atrophy in mature mice (57). These mice also exhibit pubertal testicular hypertrophy and immature germ cell accumulation nearly identical to the *Bax*-deficient mice (57). These results support a model in which overexpression of *Bcl-2* inhibits a *Bax*-dependent wave of cell death of developing spermatogonia (Fig. 14.4). However, the possibility that *Bax* also performs a novel activity required for germ cell maturation that is also blocked by binding to *Bcl-2* cannot be formally excluded by these data.

A critical issue raised by these findings is why cell death appears essential for normal testis function. One possibility is that excess germ cells need to be eliminated by apoptosis before formation of the blood–testis barrier by tight junctions between Sertoli cells (Fig. 14.4). In the absence of *Bax* or presence of excess *Bcl-2*, germ cell apoptosis does not occur and the tight junctions are unable to form. Consistent with this model, tight junctions are extremely rare in *Bax* defi-

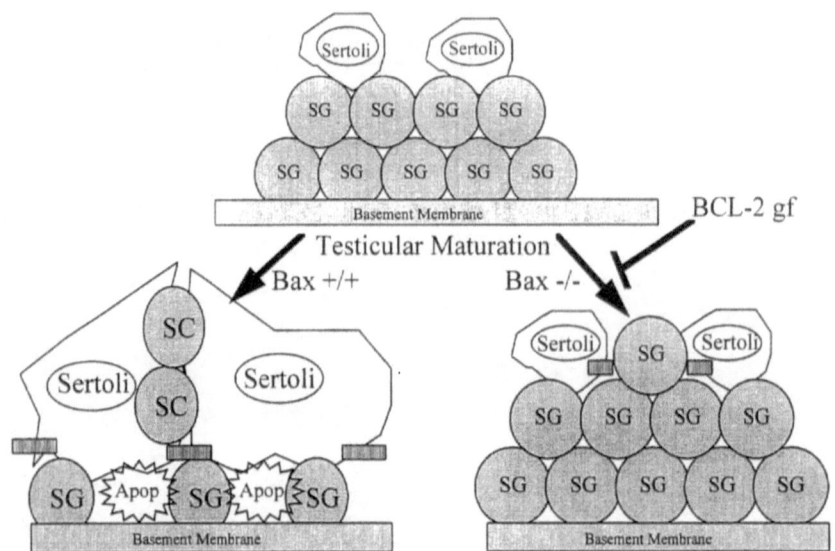

FIGURE 14.4. Model of infertility in *Bax*-deficient mice depicts the seminiferous tubule as it progresses from an immature testis (top) with spermatogonia (SG) and no tight junctions to a mature testis with tight junctions (hatched boxes) and spermatocytes (SC). In a normal animal (*Bax* +/+), some spermatogonia undergo apoptosis (Apop) and the tight junctions form. In *Bax*-deficient mice (-/-) or BCL2 gain-of-function (gf) mice, the spermatogonia do not undergo apoptosis and the Sertoli cells are unable to form tight junctions and promote maturation of spermatocytes.

cient mice (52), suggesting a model in which a *Bax*-dependent wave of spermatocyte apoptosis is critical for normal testicular development.

References

1. Huckins C. The morphology and kinetics of spermatogonial degeneration in normal adult rats: an analysis using a simplified classification of the germinal epithelium. Anat Rec 1978;190:905–26.
2. Kerr JF, Wyllie AH, Currie AR. Apoptosis: a basic biological phenomenon with wide-ranging implications in tissue kinetics. Br J Cancer 1972;26:239–57.
3. Sulston JE, Horvitz HR. Post-embryonic cell lineages of the nematode *Caenorhabditis elegans*. Dev Biol 1977;82:110–56.
4. Ellis HM, Horvitz HR. Genetic control of programmed cell death in the nematode *C. elegans*. Cell 1986;44:817–29.
5. Hengartner MO, Horvitz HR. Activation of *C. elegans* cell death protein CED-9 by an amino-acid substitution in a domain conserved in Bcl-2. Nature (Lond) 1994;369:318–20.
6. Hengartner MO, Ellis RE, Horvitz HR. *Caenorhabditis elegans* gene *ced-9* protects cells from programmed cell death. Nature (Lond) 1992;356:494–9.

7. Yuan J, Shaham S, Ledoux S, Ellis HM, Horvitz HR. The *C. elegans* cell death gene *ced-3* encodes a protein similar to mammalian interleukin-1 beta-converting enzyme. Cell 1993;75:641–52.
8. Ghayer T, Banerjee S, Hugunin M, Butler D, Herzog L, Carter A, et al. Caspase-1 precesses IFN-gamma-inducing factor and regulates LPS-induced FN-gamma production. Nature (Lond) 1997;386:619–23.
9. Alnemri ES, Livingston DJ, Nicholson DW, Salvesen GS, Thornberry NA, Wong WW, et al. Human ICE/CED-3 protease nomenclature. Cell 1996;87:171.
10. Li P, Allen H, Banerjee S, Franklin S, Herzog L, Johnston C, et al. Mice deficient in IL-1 beta-converting enzyme are defective in production of mature IL-1 beta and resistant to endotoxic shock. Cell 1995;80:401–11.
11. Kuida K, Lippke JA, Ku G, Harding MW, Livingston DJ, Su MS, et al. Altered cytokine export and apoptosis in mice deficient in interleukin-1 beta converting enzyme. Science 1995;267:2000–3.
12. Nicholson DW, Ali A, Thornberry NA, Vaillancourt JP, Ding CK, Gallant M, et al. Identification and inhibition of the Ice/ced-3 protease necessary for mammalian apoptosis. Nature (Lond) 1995;376:37–43, 109.
13. Chinnaiyan AM, Orth K, Orourke K, Duan HJ, Poirier GG, Dixit VM. Molecular ordering of the cell death pathway. Bcl-2 and Bcl-x_L function upstream of the ced-3-like apoptotic proteases. J Biol Chem 1996;271:4573–6, 13.
14. Kuida K, Zheng TS, Na SQ, Kuan CY, Yang D, Karasuyama H, et al. Decreased apoptosis in the brain and premature lethality in cpp32-deficient mice. Nature (Lond) 1996;384:368–72, 26.
15. Gagliardini V, Fernandez P-A, Lee RKK, Drexler HCA, Rotello RJ, Fishman MC, et al. Prevention of vertebrate neuronal death by the crmA gene. Science 1994;23:826–8.
16. Bump NJ, Hackett M, Hugunin M, Seshagiri S, Brady K, Chen P, et al. Inhibition of ice family proteases by baculovirus antiapoptotic protein p35. Science 1995;269:1885–8, 95.
17. Thome M, Schneider P, Hofmann K, Fickenscher H, Meinl E, Neipel F, et al. Viral FLICE-inhibitory proteins (FLIPS) prevent apoptosis induced by death receptors. Nature (Lond) 1997;386:517–21.
18. Enari M, Talanian RV, Wong WW, Nagata S. Sequential activation of ice-like and cpp32-like proteases during fas-mediated apoptosis. Nature (Lond) 1996;380:723–6, 60.
19. Xiang J, Chao DT, Korsmeyer SJ. Bax-induced cell death may not require interleukin 1β-converting enzyme-like proteases. Proc Natl Acad Sci USA 1996;93:14559–63.
20. Armstrong RC, Aja T, Xiang J, Gaur S, Krebs JF, Hoang K, et al. Fas-induced activation of the cell death-related protease CPP32 is inhibited by Bcl-2 and by ICE family protease inhibitors. J Biol Chem 1996;271:16850–5.
21. Miura M, Friedlander RM, Yuan J. Tumor necrosis factor-induced apoptosis is mediated by a CrmA-sensitive cell death pathway. Proc Natl Acad Sci USA 1995;92:8318–21.
22. McCarthy NJ, Whyte MKB, Gilbert CS, Evan GI. Inhibition of Ced-3/ICE-related proteases does not prevent cell death induced by oncogenes, DNA damage, or the Bcl-2 homologue Bak. J Cell Biol 1997;136:215–27.
23. Thornberry NA, Bull HG, Calaycay JR, Chapman KT, Howard AD, Kostura MJ, et al. A novel heterodimeric cysteine protease is required for interleukin-1 beta processing in monocytes. Nature (Lond) 1992;356:768–74.

24. Tewari M, Quan LT, Orourke K, Desnoyers S, Zeng Z, Beidler DR, et al. Yama/cpp32-beta, a mammalian homolog of ced-3, is a crma-inhibitable protease that cleaves the death substrate poly(adp-ribose) polymerase. Cell 1995;81:801–9.
25. Boldin MP, Goncharov TM, Goltsev YV, Wallach D. Involvement of MACH, a novel MORT1/FADD-interacting protease, in Fas/APO-1 and TNF receptor-induced cell death. Cell 1996;85:803–15.
26. Chinnaiyan AM, Orourke K, Lane BR, Dixit VM. Interaction of ced-4 with ced-3 and ced-9—a molecular framework for cell death. Science 1997;275:1122–6, 2.
27. Liu XS, Kim CN, Yang J, Jemmerson R, Wang XD. Induction of apoptotic program in cell-free extracts—requirement for datp and cytochrome c. Cell 1996;86:147–57, 50.
28. Yang J, Liu XS, Bhalla K, Kim CN, Ibrado AM, Cai JY, et al. Prevention of apoptosis by bcl-2—release of cytochrome c from mitochondria blocked. Science 1997;275: 1129–32, 4.
29. Kluck RM, Bossywetzel E, Green DR, Newmeyer DD. The release of cytochrome c from mitochondria—a primary site for bcl-2 regulation of apoptosis. Science 1997; 275:1132–6, 5.
30. Whyte M. ICE/CED-3 proteases in apoptosis. Trends Cell Biol 1996;6:245–8.
31. Liu X, Zou H, Slaughter C, Wang X. DFF, a heterodimeric protein that functions downstream of caspase-3 to trigger DNA fragmentation during apoptosis. Cell 1997;89:175–84.
32. Hengartner MO, Horvitz HR. *C. elegans* cell survival gene *ced-9* encodes a functional homolog of the mammalian proto-oncogene *bcl-2*. Cell 1994;76:665–76.
33. Vaux DL, Weissman IL, Kim SK. Prevention of programmed cell death in *Caenorhabditis elegans* by human bcl-2. Science 1992;258:1955–7.
34. Shimizu S, Eguchi Y, Kamiike W, Waguri S, Uchiyama Y, Matsuda H, et al. Bcl-2 blocks loss of mitochondrial membrane potential while ice inhibitors act at a different step during inhibition of death induced by respiratory chain inhibitors. Oncogene 1996;13:21–9.
35. Yang E, Korsmeyer SJ. Molecular thanatopsis: a discourse on the Bcl2 family and cell death. Blood 1996;88:386–401.
36. Reed JC. Bcl-2 and the regulation of programmed cell death. J Cell Biol 1994;124:1–6.
37. Sedlak TW, Oltvai ZN, Yang E, Wang K, Boise LH, Thompson CB, et al. Multiple Bcl-2 family members demonstrate selective dimerizations with Bax. Proc Natl Acad Sci USA 1995;92:7834–8.
38. Oltvai ZN, Korsmeyer SJ. Checkpoints of dueling dimers foil death wishes. Cell 1994;79:189–92.
39. Chittenden T, Flemington C, Houghton AB, Ebb RG, Gallo GJ, Elangovan B, et al. A conserved domain in Bak, distinct from BH1 and BH2, mediates cell death and protein binding functions. EMBO J 1995;14:5589–96.
40. Wu DY, Wallen HD, Nunez G. Interaction and regulation of subcellular localization of ced-4 by ced-9. Science 1997;275:1126–9, 3.
41. Spector MS, Desnoyers S, Hoeppner DJ, Hengartner MO. Interaction between the c-elegans cell-death regulators ced-9 and ced-4. Nature (Lond) 1997;385:653–6, 7.
42. Shaham S, Horvitz HR. Developing *Caenorhabditis elegans* neurons may contain both cell-death protective and killer activities. Genes Dev 1996;10:578–91.
43. Zamzami N, Susin SA, Marchetti P, Hirsch T, Gomez-Monterrey I, Castedo M, et al. Mitochondrial control of nuclear apoptosis. J Exp Med 1996;183:1533–44.
44. Susin SA, Zamzami N, Castedo M, Hirsch T, Marchetti P, Macho A, et al. Bcl-2

inhibits the mitochondrial release of an apoptogenic protease. J Exp Med 1996;184: 1331–41.

45. Muchmore SW, Sattler M, Liang H, Meadows RP, Harlan JE, Yoon HS, et al. X-ray and NMR structure of human Bcl-x$_L$, an inhibitor of programmed cell death. Nature (Lond) 1996;381:335–41.

46. Minn AJ, Velez P, Schendel SL, Liang H, Muchmore SW, Fesik SW, et al. Bcl-x(l) forms an ion channel in synthetic lipid membranes. Nature (Lond) 1997;385:353–7, 16.

47. Allan DJ, Harmon BV, Kerr JFR. Cell death in spermatogenesis. In: Potten CS, ed. Perspectives on mammalian cell death. Oxford: Oxford University Press, 1987:229–58.

48. Bellgrau D, Gold D, Solawry H, Moore J, Franzusoff A, Duke RC. A role for CD95 ligand in preventing graft rejection. Nature (Lond) 1995;337:630–2.

49. Adachi M, Suematsu S, Kondo T, Ogasawara J, Tanaka T, Yoshida N, et al. Targeted mutation in the Fas gene causes hyperplasia in peripheral lymphoid organs and liver. Nat Genet 1995;11:294–300.

50. Nagata S. Apoptosis by death factor. Cell 1997;88:355–65, 11.

51. Oltvai ZN, Milliman CL, Korsmeyer SJ. Bcl-2 heterodimerizes in vivo with a conserved homolog, Bax, that accelerates programmed cell death. Cell 1993;74:609–19.

52. Knudson CM, Tung KS, Tourtellotte WG, Brown GA, Korsmeyer SJ. Bax-deficient mice with lymphoid hyperplasia and male germ cell death. Science 1995;270:96–9.

53. Boise LH, Gonzalez-Garcia M, Postema CE, Ding L, Lindsten T, Turka LA, et al. bcl-x, a bcl-2-related gene that functions as a dominant regulator of apoptotic cell death. Cell 1993;74:597–608.

54. Shaham S, Horvitz HR. An alternatively spliced *C. elegans* ced-4 RNA encodes a novel cell death inhibitor. Cell 1996;86:201–8.

55. Wang L, Miura M, Bergeron L, Zhu H, Yuan J. Ich-1, an Ice/ced-3-related gene, encodes both positive and negative regulators of programmed cell death. Cell 1994;78: 739–50.

56. Veis DJ, Sorenson CM, Shutter JR, Korsmeyer SJ. Bcl-2-deficient mice demonstrate fulminant lymphoid apoptosis, polycystic kidneys, and hypopigmented hair. Cell 1993;75:229–40.

57. Furuchi T, Masuko K, Nishimune Y, Obinata M, Matsui Y. Inhibition of testicular germ cell apoptosis and differentiation in mice misexpressing bcl2 in spermatogonia. Development (Camb) 1996;122:1703–9.

15

Hormonal Regulation of Germ Cell Apoptosis

Ronald S. Swerdloff, YanHe Lue, Christina Wang,
Tripathi Rajavashisth, and Amiya Sinha Hikim

Apoptosis is a genetically encoded programmed cell death mechanism in which cells die in a controlled fashion either spontaneously or in response to changes in the levels of specific physiological stimuli (1–4). This unique mode of cell death has long been known to be a fundamental feature of animal development and serves as a prominent mechanism in the elimination of cells that have been produced in excess, which have developed improperly, or that have sustained genetic damage, and can be modulated in various cell types by a wide variety of regulatory stimuli (5). These diverse regulatory stimuli or signals may act to either suppress or promote the activation of the death program, and the same signals may actually have opposing effects on different cell types. Derangements of apoptosis can have deleterious consequences, as noted in several human disease states including cancer, AIDS (acquired immunodeficiency syndrome), autoimmuno-diseases, and neurodegenerative disorders (5–7).

Spermatogenesis is a complex process in which immature, undifferentiated spermatogonia divide and differentiate into mature spermatozoa. This is a highly organized, precisely timed process composed of a repetitive cycle (the seminiferous epithelial cycle) that can be subdivided into a number of stages (8, 9). In rats, there are 14 stages and each stage is characterized by a unique association of germ cells that are at different levels of their differentiation into spermatozoa. It has long been recognized that, in normal adult rats, there are three critical points in differentiation of germ cells at which extensive cell loss occurs (8–11). The first is during spermatogonial development and involves primarily types A_2 and A_3 spermatogonia; the second is during the meiotic divisions; and the last is during the acrosome and maturation phases of spermiogenesis. In summation, germ cell death during normal spermatogenesis has been estimated to result in loss of as much as 75% of potential numbers of mature sperm cells in the adult testis (10).

Withdrawal of gonadotropins or testosterone (T) further enhances the degeneration of germ cells in the testis (12–14).

Extending these earlier morphological analysis, recent studies have demonstrated that programmed cell death or apoptosis is the underlying mechanism of germ cell death (15–19). These studies suggest that apoptosis is an important determinant of hormone-dependent survival of germ cells in the testis. Our focus in this review is on the hormonal regulation of germ cell apoptosis in the adult. We examine the involvement of apoptosis in the induction of germ cell degeneration in adult rats after selective suppression of gonadotropin by gonadotropin-releasing hormone antagonist (GnRH-A) treatment. We address the specific role and relative contribution of follicle-stimulating hormone (FSH) and luteinizing hormone (LH) in preventing GnRH-A-induced germ cell apoptosis. We examine the underlying mechanism of germ cell death after mild testicular hyperthermia. Evidence is also presented demonstrating the role of a tumor suppressor gene (p53) and a cytokine (macrophage-colony stimulating factor, M-CSF) in regulating germ cell survival.

Involvement of Apoptosis in the Induction of Germ Cell Degeneration in Adult Rats After GnRH-A Treatment

We have previously reported that selective deprivation of gonadotropin and testicular T by GnRH-A treatment is followed by stage-specific degeneration of germ cells, which are at different phases of their differentiation into spermatozoa (14). The rapid induction of selective germ cell degeneration after GnRH-A treatment thus provides an excellent in vivo model system for studying the underlying mechanism of germ cell death in the testis. Accordingly, we sought to examine, using this model system, the degree to which spontaneous and induced (in response to lack of survival factors) germ cell degeneration in the adult rat occurs by apoptosis (17).

Adult male Sprague-Dawley (SD) rats were given either a daily injection of Nal-Glu-GnRH-A (1.25 mg/kg bw) for 0 (control), 2, or 5 days. The onset of germ cell degeneration was assessed by high-resolution light and electron microscopy, as well as by a germ cell degeneration assay. The occurrence of apoptosis was characterized by (i) detection of internucleosomal DNA fragmentation after agarose gel electrophoresis, and (ii) direct immunoperoxidase detection of digoxigenin-labeled genomic DNA in specific cell types.

The earliest morphological response to gonadotropin withdrawal was noted by day 5, with the occurrence of degeneration of preleptotene and pachytene spermatocytes and step 7 spermatids at stage VII and of step 19 spermatids at stages VII and VIII of the seminiferous epithelial cycle. The onset of germ cell degeneration was accompanied by a significant ($p < .05$) increase (2.4-fold) in the degree of low molecular weight DNA fragmentation, indicating an intimate association between apoptosis and germ cell degeneration. These results have been further substantiated by in situ detection of germ cell apoptosis. Spontaneous apoptosis of differentiating spermatogonia, the type A spermatogonial population in particular, during their intensive period of proliferation was readily observed in the control animals. Immunostaining of advanced germ cells, other than a few spermatocytes during their meiotic divisions at stage XIV, was rarely observed in control rats. In contrast, intense immunoreactivity for low molecular weight DNA fragmentation was observed in the nuclei of those germ cells (preleptotene and pachytene spermatocytes, and step 7 and 19 spermatids) exhibiting morphological signs of degeneration (stage VII of the seminiferous cycle) 5 days after GnRH-A treatment. The apoptotic mode of spontaneous germ cell death and the stage-specific activation of apoptosis in response to hormonal deprivation in the adult rat as noted in this study are further supported by recent studies using GnRH-A (19), ethane dimethanesulfonate (EDS) (18) or estradiol-treated rat model (20). These results, however, do not agree with those of Billig et al. (16), who reported no changes in level of apoptotic DNA fragmentation in the adult rat after GnRH-A treatment in spite of marked suppression of plasma LH and FSH. Because our data are so clearly demonstrative, it appears that this discrepancy most likely reflects the short treatment duration (4 days) or lack of complete suppression of plasma LH levels, as suggested by those authors.

It is uncertain why some germ cells respond more readily (by undergoing apoptosis) than others to a lack of hormonal stimulation. However, the simultaneous activation of cell death involving preleptotene and pachytene spermatocytes and step 7 and step 19 spermatids, which are at very different positions in the hierarchy of germ cell lineage, may indicate that the local environment provided and controlled by the Sertoli cells is particularly critical at this time point of the cycle (9, 21). Because FSH and androgen receptors are found exclusively in the Sertoli cells, it is likely that these hormones exert their stimulatory effects on Sertoli cells, which in turn results in stimulation of intratubular factors(s) essential for the survival of germ cells through a paracrine mechanism. It is of particular interest that, within 2 days after GnRH-A treatment, there is a dramatic loss of T production, and this precedes the initiation of apoptosis in the germ cells. There is growing evidence to suggest that stages VII–VIII of the rat spermatogenic cycle exhibit the highest levels of immunocytochemically detectable androgen receptor expression (22) and are considered to be classically androgen dependent (9, 23). Thus, the marked reduction of intratesticular T concentration appears to be an important initiator of programmed cell death in the seminiferous epithelium.

To explore the possibility that apoptosis is the sole mechanism of germ cell loss in the adult rat after selective withdrawal of gonadotropins and T, we first investigated the temporal and stage-specific activation of germ cell apoptosis in rats treated daily with GnRH-A for as long as 2 weeks. We then examined the relationship between germ cell degeneration (characterized by morphological criteria) and apoptosis. Groups of adult male SD rats were given a daily injection of a vehicle for 14 days or GnRH-A (1.25 mg/kg bw) for 2, 5, 7, and 14 days. The occurrence of apoptosis was characterized by genomic DNA electrophoresis and in situ 3′-end-labeling histochemistry in glutaraldehyde-fixed, paraffin-embedded sections. Glutaraldehyde fixation was used (i) to improve both the specificity and the sensitivity of in situ 3′-end-labeling of apoptotic DNA fragmentation and (ii) to maintain the excellent morphological preservation needed to resolve stage-related susceptibility of specific germ cells to programmed cell death. Enumeration of viable Sertoli nuclei having distinct nucleoli and the apoptotic germ cell population was carried out at stages I, II–IV, V–VI, VII–VIII, IX–XI, and XII–XIV. The rate of germ cell apoptosis (apoptotic index) was expressed as the number of apoptotic germ cells per Sertoli cell. The rate of germ cell degeneration (degeneration index) was determined employing the same approach used to determine the germ cell apoptotic index. Figure 15.1 shows representative examples of in situ 3′-end-labeling of DNA strand breaks in the presence (A) or absence (B) of a counterstain in germ cells undergoing apoptosis in response to early deprivation of gonadotropins. As shown in Figure 15.2, analysis of testicular apoptotic DNA fragmentation revealed a detectable increase at day 5 and a maximal increase at 14 days after treatment. In situ analysis of germ cell apoptosis fully corroborated the observed increase in the degree of DNA fragmentation with time and also revealed a stage-related activation of apoptosis of specific germ cells (Table 15.1). A low incidence of germ cell apoptosis was detectable at stages I, IX–XI, and XII–XIV in control rats. The mean number of apoptotic germ cells, specifically at stages VII–VIII, increased significantly by day 5 and increased another 2.2 fold (over the 5-day treatment values) on day 7 after GnRH-A treatment when compared to controls, in which no apoptosis was detected. Also, significantly increased apoptosis at stages IX–XI over controls was noted by day 7. Maximal increases in the number of dying cells occurred by day 14, at which time an increased incidence of apoptosis was also noted at stages I, II–IV, V–VI, and XII–XIV in comparison with controls. Stages VII–VIII, and IX–XI, however, still exhibited the highest number of germ cells undergoing apoptosis. A comparison between the rates of apoptosis and cell degeneration measured at stages VII–VIII demonstrated an intimate association ($r = .94$; $p < .001$) between apoptosis and overall germ cell loss.

The results of this study confirm and extend earlier reports by demonstrating a stage-related susceptibility of germ cells to apoptosis after acute withdrawal of gonadotropins or T (17–19). Stages VII–VIII, followed later by stages IX–XI, were the first to show accelerated germ cell apoptosis between 5 and 7 days. These stages also exhibited the highest number of germ cells undergoing apoptosis by day 14. However, by this time the late (XII–XIV) and early (I–VI) stages also exhibited a modest but significant increase in the number of apoptotic germ cells. It is most likely that the accelerated germ cell apoptosis seen at later stages after 2 weeks of GnRH-A treatment is a delayed consequence of abnormal development of germ cells as they pass through stages VII–VIII. However, it should be noted here that as more and more germ cells undergo apoptosis at stages VII–XI, then fewer germ cells pass into stages I–VI and eventually again through stages VII–VIII of the next cycle. Thus, it can be envisaged how spermatogenesis gradually and progressively winds down in response to a lack of hormonal stimulation.

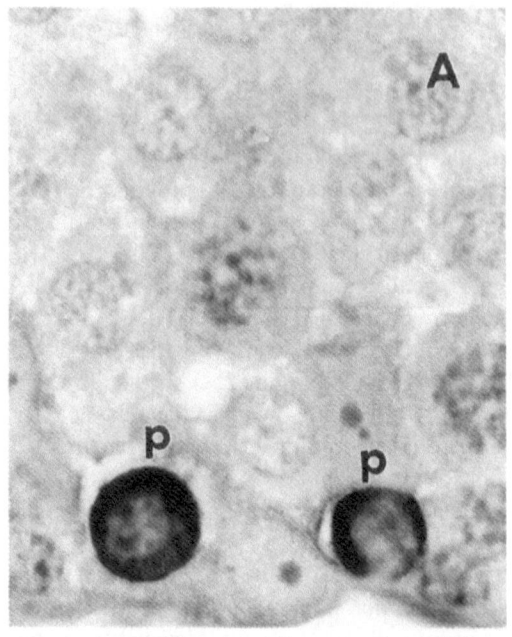

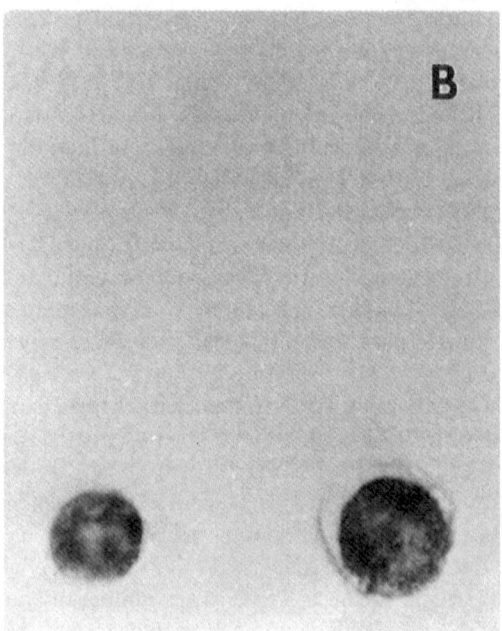

FIGURE 15.1. In situ 3′-end-labeling of DNA strand breaks in apoptotic germ cells in glutaraldehyde-fixed, paraffin-embedded testicular sections. (*A*) Portion of a stage VII tubule from a rat treated with gonadotropin-releasing hormone antagonist (GnrH-A) for 5 days exhibiting two apoptotic pachytene (P) spermatocytes. Methyl green counterstain. Note intense labeling along the nuclear periphery. (*B*) Representative examples of immunostaining detected in the absence of a counterstain in germ cells undergoing apoptosis. ×1400.

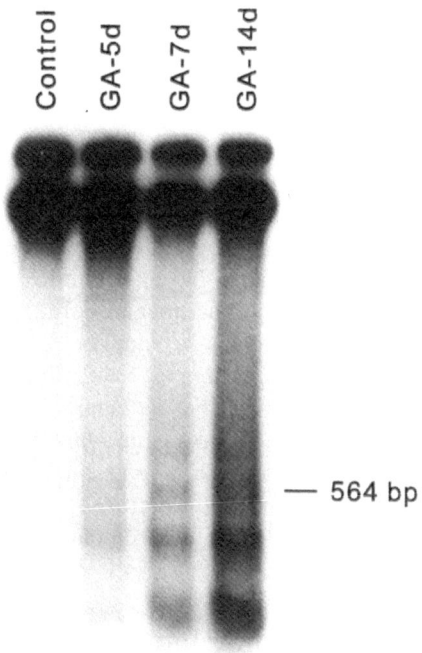

**GnRH Antagonist (GA)
Treatment**

FIGURE 15.2. GnRH-A-induced (GA) testicular apoptotic DNA fragmentation. Low molecular weight DNA fragmentation could be detected as early as 5 days, and became more pronounced between 7 and 14 days after GnRH-A treatment. From Sinha Hikim et al., unpublished observation.

TABLE 15.1. Quantitative assessment of apoptotic germ cell population (expressed as number/ Sertoli cell) at designated stages of the seminiferous epithelial cycle after gonadotropin-releasing hormone antagonist (GnRH-A) (GA) treatment.

Treatment	Seminiferous epithelial stages					
	I	II–IV	V–VI	VII–VIII	IX–XI	XII–XIV
Control	0.06 ± 0.01[a]	0[a]	0[a]	0[a]	0.07 ± 0.01[a]	0.09 ± 0.02[a]
GA-2d	0.07 ± 0.01[a]	0[a]	0[a]	0.03 ± 0.01[a]	0.06 ± 0.02[a]	0.08 ± 0.01[a]
GA-5d	0.06 ± 0.01[a]	0[a]	0[a]	0.40 ± 0.06[b]	0.08 ± 0.02[a]	0.09 ± 0.02[a]
GA-7d	0.08 ± 0.01[a]	0[a]	0[a]	0.86 ± 0.04[c]	0.37 ± 0.05[b]	0.10 ± 0.01[a]
GA-14d	0.34 ± 0.03[b]	0.16 ± 0.07[b]	0.40 ± 0.19[b]	0.97 ± 0.07[c]	1.03 ± 0.22[c]	0.24 ± 0.01[b]

From Sinha Hikim et al., unpublished data.
Values are the mean $\pm$ SEM. In each column, means with unlike superscripts (a–c) are significantly ($p < .05$) different.

These results, together with previous findings that GnRH-A treatment induces azoospermia in the rat (14, 24), suggest that apoptosis plays a major role in progressive and organized regression of spermatogenesis after gonadotropin deprivation. Further support for this suggestion stems from the earlier findings that normal proliferation of the differentiating spermatogonia was not affected by gonadotropin deprivation (14, 25, 26). The possibility that premature loss of round spermatids by detachment may have contributed to the observed suppression of spermatogenesis as reported in other model systems (27, 28) seems most unlikely, because we did not see such sloughing of germ cells in this study or in our earlier studies (14, 24, 25), even after 4 weeks of GnRH-A treatment when the rat became completely azoospermic. Experimental hormonal male contraceptive approaches have utilized various agents to suppress LH and FSH levels and induce azoospermia and severe oligospermia. Based on our experiences with hypophysectomy and GnRH antagonist-induced suppression of LH and FSH in rodents, it is our belief that apoptosis is also an important determinant of hormonal male contraceptive-induced spermatogenic suppression in men. We do know that spontaneous germ cell degeneration during normal spermatogenesis in men involves apoptosis (Sinha Hikim et al., unpublished data).

Possible Role of Recombinant Human (rh)LH and FSH in the Prevention of GnRH-A-Induced Apoptosis of Germ Cells in Adult Rats

To understand the hormonal regulation of germ cell apoptosis, we then examined the extent to which rhFSH or rhLH is able to prevent GnRH-A-induced germ cell apoptosis in the adult rat. Adult male rats were given a daily injection of vehicle (controls), GnRH-A (1.25 mg/kg bw), GnRH-A + 2.5 or 10 IU rhLH, or GnRH-A + 10 IU FSH for 7 days. Enumeration of the apoptotic germ cell population (characterized by in situ end-labeling histochemistry) was carried out at stages I, II–IV, V–VI, VII–VIII, IX–XI, and XII–XIV of the spermatogenic cycle. FSH dose was based on the results of our previous study (25), which showed that a daily dosage of 10 IU rhFSH was capable of delaying GnRH-A-induced regression of spermatogenesis (based on morphological findings). LH doses are based on the results of our dose–response study, which showed that concurrent administration of 2.5 IU rhLH for 1 week prevented GnRH-A induced decline in testicular weight and in the numbers of homogenization-resistant advanced spermatids and significantly increased both the concentration (48.1% of controls) as well as the content (47.7% of controls) of testicular T, but had no effect on plasma levels of T. However, at slightly higher doses (10 IU), rhLH resulted in a marked increase (2.2 fold over control levels) in testicular T levels and normalized the levels of circulating T.

Concomitant administration of either 2.5 or 10 IU rhLH for 1 week attenuated the GnRH-A-induced germ cell apoptosis (Table 15.2). However, it should be

TABLE 15.2. Gonadotropin suppression of GnRH-A-induced stage-specific activation of germ cell apoptosis.

Stages	Apoptotic germ cells (number/Sertoli cell)				
	Control	GnRH-A	GnRH-A + 2.5 IU LH	GnRH-A + 10 IU LH	GnRH-A + 10 IU FSH
I	0.08 ± 0.01^a	0.05 ± 0.03^a	0.10 ± 0.03^a	0.11 ± 0.05^a	0.13 ± 0.03
II–IV	0	0	0	0	0
IV–VI	0	0	0	0	0
VII–VIII	0^a	0.75 ± 0.14^b	0.10 ± 0.003^a	0.005 ± 0.003^a	0.28 ± 0.06^c
IX–XI	0.04 ± 0.03^a	0.46 ± 0.18^b	0.09 ± 0.003^a	0.09 ± 0.02^a	0.12 ± 0.03^a
XII–XIV	0.09 ± 0.02^a	0.11 ± 0.01^a	0.09 ± 0.02^a	0.11 ± 0.02^a	0.08 ± 0.01^a

From Sinha Hikim et al., unpublished data.
LH, luteinizing hormone; FSH, follicle-stimulating hormone.
Values are the mean ± SEM. In each row, means with unlike superscripts (a–c) are significantly ($p < .05$) different.

noted here that previous studies on GnRH-A-treated rats have shown that exogenous administration of T increases both the serum concentrations and pituitary content of FSH as well as maintaining spermatogenesis (29, 30). It is, therefore, possible that GnRH-A-treated animals given LH, which results in increased plasma T concentrations, will secondarily lead to an increase in plasma FSH levels. To pursue this possibility, we further assessed the changes in plasma FSH levels in GnRH-A-treated rats in response to increasing doses (2.5–30.0 IU) of rhLH. Concurrent administration of LH at the lowest dose level in GnRH-A-treated rats did not increase FSH levels above that of GnRH-A-treated rats alone (data not shown). These results further demonstrate that the observed stimulatory effects of rhLH at the lowest dose level (2.5 IU) on spermatogenesis are specifically caused by LH alone. Treatment with FSH also effectively suppressed the GnRH-A-induced germ cell apoptosis at stages VII–VIII by 67% and stages IX–XI by 74%.

Analysis of testicular internucleosomal DNA fragmentation, using a sensitive 3'-end DNA-labeling technique followed by size fractionation in agarose gels and autoradiography, further confirmed the preventive effect of either gonadotropin on GnRH-A-induced germ cell apoptosis (data not shown). This situation is very different from the immature rat, where human chorionic gonadotropin (hCG) (LH surrogate) or T was unable to fully prevent the hypophysectomy-induced germ cell apoptosis, even at earlier time (4 days) intervals (15). It is pertinent to note here that the maturation of Sertoli cells involves a progressive switch from being mainly FSH modulated in immature animals to being mainly T modulated in the adult (9). This developmental switch in hormonal dependence of Sertoli cell function presumably also explains why LH was able to fully prevent the GnRH-A-induced germ cell apoptosis in adult rats but failed to do so in immature animals. However, it is possible that the stimulatory effects of either gonadotropin that are obvious after 1 week of gonadotropin and T deprivation might not become so obvious after long-term treatment. The precise role and relative contribution of these hormones either alone or in combination in determining the fate of specific germ cells are also not known. These important issues are currently under investigation in our laboratory. Collectively, these results suggest that (i) pituitary gonadotropins LH and, to a lesser extent, FSH are the important extrinsic regulators of germ cell apoptosis, and (ii) hormonal regulation of germ cells between immature and adult rats is decisively different.

Germ Cell Apoptosis After Mild Testicular Hyperthermia

Classical histological studies have shown that local heating of the testis in rats (43°C for 15 min) results in accelerated death of germ cells involving spermatocytes and early spermatids (11, 31, 32). However, the underlying mechanism of heat-induced germ cell death remains poorly understood. Our major objectives herein were (i) to examine whether the heat-induced loss of germ cells in the adult rat occurs by accelerated apoptosis; (ii) if so, to characterize the specific germ cell types that are most sensitive to mild testicular hyperthermia; and (iii) to analyze the temporal and stage-specific activation of germ cell apoptosis. The scrotal regions of groups of five adult male rats were immersed in a water bath at room temperature 22°C (control) or at 43°C for 15 min. Animals were then killed on day 1, 2, or 9 after heat exposure. Apoptosis was characterized by a modified in situ procedure as described and quantitated as number of apoptotic germ cell per Sertoli cell (apoptotic index). As shown in Fig. 15.3, mild hyperthermia within 1 or 2 days resulted in a marked activation of germ cell apoptosis specifically at

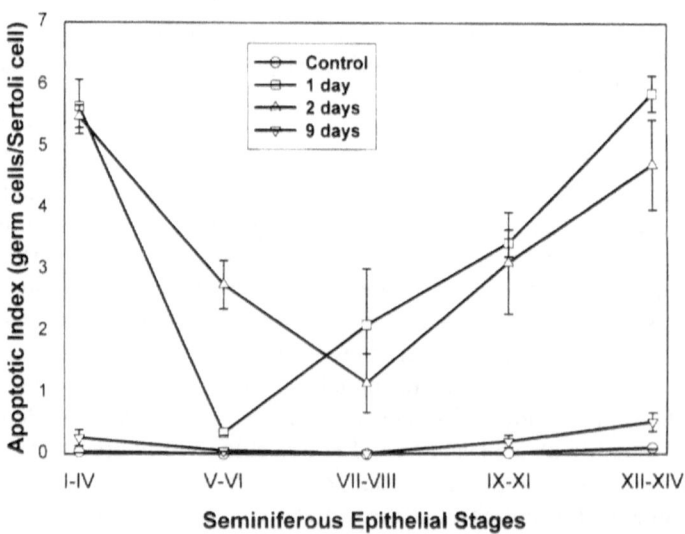

FIGURE 15.3. Stage-specific activation of germ cell apoptosis in rat after mild testicular hyperthermia. From Lue et al., unpublished observation.

early (I–IV) and late (XII–XIV) stages. A mild reaction was shown, however, by the tubules at stages V–VI and VII–VIII. By day 9, most of the tubules were severely damaged and displayed only a few remaining apoptotic germ cells because most of these dead cells were eliminated (presumably) through phagocytosis by the Sertoli cells. Of interest, the effect of heat on spermatogenesis is not only stage specific but also cell specific (Fig. 15.4). Spermatocytes, including

Stages	I-IV	V-VI	VII-VIII	IX-XI	XII-XIV
Duration (hours)	79	56	88	21	64

Day 1

15-17	17-18	19		
1-4	5-6	7-8	9-11	12-14
P	P	P	P	P, D, DV
IN	B	PL	L	Z
A	A	A	A	A

Day 2

15-17	17-18	19		
1-4	5-6	7-8	9-11	12-14
P	P	P	P	P, D, DV
IN	B	PL	L	Z
A	A	A	A	A

Day 9

15-17	17-18	19		
1-4	5-6	7-8	9-11	12-14
P	P	P	P	P, D, DV
IN	B	PL	L	Z
A	A	A	A	A

FIGURE 15.4. Heat-induced stage-related activation of apoptosis inolving specific germ cells. Shaded areas represent the cell types that show a very high incidence of apoptosis or are missing (presumably by phagocytosis) at indicated time intervals after a single exposure to heat (43°C for 15 min). A, type A spermatogonia; IN, intermediate spermatogonia; B, type B spermatogonia; PL, preleptotene; L, leptotene; Z, zygotene; P, pachytene; D, diplotene; DV, dividing spermatocytes. Numbers 1–19 refer to the successive steps in spermatid development. From Lue et al., unpublished observations.

pachytenes at stages I–V and IX–XII, diplotene and dividing spermatocytes at stages XIII–XIV, and early (steps 1–4) spermatids were most susceptible to heat. Differentiating spermatogonia and the Sertoli cells, however, remained unaffected. These results suggest that (i) the adverse effects of mild testicular hyperthermia on spermatogenesis are mediated by a stage-related activation of apoptosis involving specific germ cells and (ii) early (I–IV) and late (XII–XIV) stages, but not the hormone-sensitive stages (VII–VIII), are most sensitive to heat. Furthermore, these results also suggest that physical stimulus, such as mild heat, can induce germ cell apoptosis possibly through different pathways than those involved in the activation of apoptosis by gonadotropin deprivation.

Genetic Modulation of Germ Cell Apoptosis

A recent and exciting advance in understanding the genetic modulation of apoptosis is the utilization of genetically altered mice either overexpressing or harboring a null mutation of specific genes. Studies using these mice are making an increasing contribution to our understanding of the role of various genes in regulating apoptosis in various systems. These animals would also be valuable in studying the in vivo functions of various targeted genes in regulating the fate of germ cells during the normal spermatogenic process or in response to a lack of hormonal stimulation.

Suppression of Spontaneous Germ Cell Apoptosis in p53-Deficient Mice

Mutations of the p53 gene are the most frequent abnormalities in various human tumors (33, 34). Available data from studies in extragonadal cell systems suggest that p53 is an important regulator of both cell proliferation (via its ability to halt cell-cycle progression) and apoptosis (33, 34). In this study, using p53-deficient mice, we sought to determine the role of p53 in spontaneous germ cell apoptosis. Normal and p53-deficient (on a C57BL/6 genetic background) adult male mice were obtained from the Jackson Laboratory (Bar Harbor, ME, USA). Apoptosis was characterized by in situ end-labeling histochemistry using glutaraldehyde-fixed testicular sections as described. In situ analyses revealed a stage-specific occurrence of germ cell apoptosis in normal mice. The lowest level (4.6–5.5) of apoptosis (expressed as number of apoptotic cells/100 Sertoli cells) was detected at stages VII–VIII and IX–XI and the highest levels (17.7–28.2) in stages I–III, IV–VI, and XII. The level of spontaneous apoptosis, specifically at stages I–III and XII of p53-deficient mice, was significantly ($p < .02$) lower than those of the normal animals. Pachytene, as well as the dividing spermatocytes in late meiosis, were the major cell types rescued. Lower incidence of apoptosis was further reflected by a significantly ($p < .02$) higher testicular sperm count in p53-deficient mice ($27.6 \pm 0.5 \times 10^6$) compared to normal mice ($22.8 \pm 1.4 \times 10^6$). Serum and testicular testosterone levels were somewhat higher in p53-deficient males than in normal mice, but these differences were not statistically significant. These results suggest a specific role of p53 in regulating germ cell apoptosis during normal spermatogenesis. p53 inactivation enhances germ cell survival by inhibiting germ cell apoptosis predominantly at nonhormonally protected stages.

Stage-Specific Activation of Germ Cell Apoptosis in Mice Lacking Macrophage-Colony Stimulating Factor

Osteopetrotic (op/op) mice lacking M-CSF as a result of a structural gene mutation have impaired fertility in both males and females (35, 36) and exhibit a phenotype that includes severe deficiency in blood monocytes and in peritoneal as well as tissue macrophages (37). There is also a rapidly growing appreciation regarding the roles of macrophages in fine-tuning testicular function, including steroidogenesis and spermatogenesis (37). In op/op mice, both serum and testicular T levels were markedly reduced ($p < .01$) as compared to heterozygous (op/+) or normal littermates. In situ analysis of germ cell apoptosis, characterized by 3'-end DNA labeling coupled with immunocytochemistry, in these mice revealed an increased number of apoptotic germ cells in a stage-dependent manner. The apoptotic indices (number of dying germ cells per 100 Sertoli cells) were significantly higher in stages IV–VI and XII, as compared to other stages in mutant mice and to the corresponding stages in normal mice. These results provide direct evidence in support of the previous concept, that the macrophage plays a role in spermatogenesis and in testicular steroidogenesis. M-CSF null mutation affects germ cell survival by accelerating germ cell apoptosis at nonhormonally protected stages.

Summary

Both spontaneous and accelerated germ cell death triggered by deprivation of gonadotropic support by GnRH-A treatment in the adult rat occur almost exclusively via apoptosis. Apoptosis in germ cells is not random but is highly selective and occurs after different interventions in specific stages of the seminiferous epithelial cycle. Stages VII–VIII, followed later by stages IX–XI, were the first to show accelerated germ cell death after acute (between 5 and 7 days) withdrawal of gonadotropins. Maximal increase in the number of dying cells occurred by day 14, at which time the late (XII–XIV) and early (I–VI) stages also exhibited a modest increase in the number of apoptotic germ cells. It is most likely that the accelerated germ cell apoptosis seen at later stages after 2 weeks of GnRH-A treatment is a delayed consequence of apoptotic programming as germ cells pass through stages VII–VIII. These results strongly support the hypothesis that apoptosis is predominantly responsible for the progressive and organized regression of spermatogenesis to azoospermia after gonadotropin deprivation. Germ cell apoptosis occurring after GnRH-A treatment can be reduced by concomitant administration of either FSH or LH. In the adult rat, LH inhibition of accelerated apoptosis was complete while FSH only partially attenuated the GnRH-A induced germ cell apoptosis. This situation is very different from the immature rat, where hCG (LH surrogate) and T were not as effective as FSH in preventing hypophysectomy-induced germ cell apoptosis.

Physical factors accelerate stage-specific apoptosis differently than hormonal deprivation. Heating the testis of adult male rats to 43°C for 15 min resulted in marked acceleration of apoptosis by day 2 at stages I–IV, IX–XI, and XII–XIV. The hormone-sensitive stages VII–VIII as well as V–VI were relatively spared. In additional studies using genetically altered mice models, we have characterized the in vivo function of a tumor suppressor gene (p53) and a cytokine (M-CSF) in regulating spontaneous germ cell apoptosis. Mice lacking p53 exhibit a stage-specific suppression of spontaneous germ cell apoptosis involving stages I–III, IV–VI, and XII. In contrast, osteopetrotic mice (*op/op*) lacking M-CSF show accelerated germ cell apoptosis at stages IV–VI and XII. These results suggest that germ cell apoptosis is a hormonally regulated process and is modulated by many extrinsic and intrinsic factors, including cytokines and tumor suppressor genes. Accelerated germ cell apoptosis may be an important mechanism involved in male infertility and a targeted area for male contraceptive development.

Acknowledgment. The authors gratefully acknowledge Dr. James S. Hutchison, Ares Advanced Technology, Inc., Randolph, MA (the Ares-Serono Group) for his generous gift of the rhLH and rhFSH.

References

1. Wyllie AH, Kerr JFR, Currie AR. Cell death: the significance of apoptosis. Int Rev Cytol 1980;68:251–306.
2. Schwartzman RA, Cidlowski JA. Apoptosis: the biochemistry and molecular biology of programmed cell death. Endocr Rev 1993;14:133–51.
3. Majno G, Joris I. Apoptosis, oncosis, and necrosis—an overview of cell death. Am J Pathol 1995;146:3–15.
4. Samali A, Gorman AM, Cotter TG. Apoptosis—the story so far. Experientia (Basal) 1996;52:933–41.
5. Thompson CB. Apoptosis in the pathogenesis and treatment of disease. Science 1995;267:1456–62.
6. Savil J. Apoptosis in disease. Eur J Clin Invest 1994;24:715–23.
7. Nagata S, Golstein P. The Fas death factor. Science 1995;267:1449–55.
8. Russell LD, Etllin RA, Sinha Hikim AP, Clegg ED. Histological and histopathological evaluation of the testis. Clearwater, FL: Cache River Press, 1990.
9. Sharpe RM. Regulation of spermatogenesis. In: Knobil E, Neil JD, eds. The physiology of reproduction. New York: Raven Press, 1994:1363–434.
10. Huckins C. The morphology and kinetics of spermatogonial degeneration in normal adult rats: an analysis using a simplified classification of germinal epithelium. Anat Rec 1978;190:905–26.
11. Allan DJ, Harmon BV, Kerr JFR. Cell death in spermatogenesis. In: Potten CS, ed.

Perspectives on mammalian cell death. London: Oxford University Press, 1987:229–58.

12. Russell LD, Clermont Y. Degeneration of germ cells in normal, hypophysectomized and hormone treated hypophysectomized rats. Anat Rec 1997;187:347–66.

13. Bartlett JMS, Kerr JB, Sharpe RM. The effect of selective destruction and regeneration of rat Leydig cells on the intratesticular distribution of testosterone and morphology of the seminiferous epithelium. J Androl 1986;7:240–53.

14. Sinha Hikim AP, Swerdloff RS. Temporal and stage-specific changes in spermatogenesis of rat after gonadotropin deprivation by a potent gonadotropin-releasing hormone antagonist treatment. Endocrinology 1993;133:2161–70.

15. Tapanainen JS, Tilly JL, Vihko KK, Hsueh AJW. Hormonal control of apoptotic cell death in the testis: gonadotropins and androgens as testicular cell survival factors. Mol Endocrinol 1993;7:643–50.

16. Billig H, Furata I, Rivier C, Tapanainen J, Parvinen M, Hsueh AJW. Apoptosis in testis germ cells: developmental changes in gonadotropin dependence and localization to selective stages. Endocrinology 1995;136:5–12.

17. Sinha Hikim AP, Wang C, Leung A, Swerdloff RS. Involvement of apoptosis in the induction of germ cell degeneration in adult rats after gonadotropin-releasing hormone antagonist treatment. Endocrinology 1995;136:2770–5.

18. Henriksen K, Hakovirta H, Parvinen M. Testosterone inhibits and induces apoptosis in rat seminiferous tubules in a stage-specific manner: *in situ* quantification in squash preparations after administration of ethane dimethane sulfonate. Endocrinology 1995;136:3285–91.

19. Brinkworth MH, Weinbauer GF, Schlatt S, Nieschlag E. Identification of male germ cells undergoing apoptosis in adult rats. J Reprod Fertil 1995;105:25–33.

20. Blanco-Rodriguez J, Martinez-Garcia. Induction of apoptotic cell death in seminiferous tubule of the adult rat testis: assessment of the germ cell types that exhibit the ability to enter apoptosis after hormone suppression by oestradiol treatment. Int J Androl 1996;19:237–47.

21. Bardin CW, Cheng CY, Musto NA, Gunsalus G. The Sertoli cell. In: Knobil E, Neil JD, eds. The physiology of reproduction. New York: Raven Press, 1994:1363–434.

22. Bremner WJ, Millar MR, Sharpe RM, Saunders PTK. Immunohistochemical localization of androgen receptors in the rat testis: evidence for stage-dependent expression and regulation by androgen. Endocrinology 1994;135:1227–34.

23. Parvinen M. Cyclic function of Sertoli cells. In: Russell LD, Griswold MD, eds. The Sertoli cell. Clearwater, FL: Cache River Press, 1993:331–47.

24. Sinha Hikim AP, Swerdloff RS. Time course of recovery of spermatogenesis and Leydig cell function after cessation of gonadotropin-releasing hormone antagonist treatment in the adult rat. Endocrinology 1994;134:1627–34.

25. Sinha Hikim AP, Swerdloff RSS. Temporal and stage-specific effects of recombinant human follicle-stimulating hormone on the maintenance of spermatogenesis in gonadotropin-releasing hormone antagonist-treated rat. Endocrinology 1995;136:253–61.

26. McLachlan RI, Wreford NG, DeKretser DM, Robertson DM. The effects of recombinant follicle-stimulating hormone on the restoration of spermatogenesis in the gonadotropin-releasing hormone-immunized adult rat. Endocrinology 1995;136:4035–43.

27. O'Donnel L, McLachlan RI, Wreford NG, Robertson DM. Testosterone promotes the conversion of round spermatids between stages VII and VIII of the rat spermatogenic cycle. Endocrinology 1994;135:2608–14.

28. McLachlan RI, Wreford NG, Meachem SJ, de Kretser DM, Robertson DM. Effects of testosterone on spermatogenic cell populations in the adult rat. Biol Reprod 1994;51: 945–55.
29. Rea MA, Marshall GR, Weinbauer GF, Nieschlag E. Testosterone maintains pituitary and serum FSH and spermatogenesis in GnRH-A suppressed rats. J Endocrinol 1986;108:101–7.
30. Bhasin S, Fielder T, Peacock N, Sod-Moriah A, Swerdloff R.S. Dissociating antifertility effects of GnRH-A from its adverse effects on mating behavior in male rats. Am J Physiol 1988;254:E84–91.
31. Chowdhury AK, Steinberger E. A quantitative study of the effect of heat on germinal epithelium of rat testes. Am J Anat 1964;115:509–24.
32. Collins P. Lacy D. Studies on the structure and function of the mammalian testis. II. Cytological and histochemical observations on the testis of the rat after a single exposure to heat applied for different lengths of time. Proc Soc Lond 1969;172:17–38.
33. Yonish-Rouach E. The p53 tumor suppressor gene: a mediator of G_1 arrest and apoptosis. Experientia (Basel) 1996;52:1001–7.
34. Levine AJ. p53, the cellular gatekeeper for growth and division. Cell 1997;88:323–31.
35. Pollard JW, Hunt JW, Wiktor-Jedrzejczak W, Stanley ER. A pregnancy defect in the osteopetrotic (op/op) mouse demonstrates the requirement for CSF-1 in female fertility. Dev Biol 1991;148:273–82.
36. Cohen PE, Chisholm O, Arceci RJ, Stanley ER, Pollard JW. Absence of colony-stimulating factor-1 in osteopetrotic (csfm^{op}/csfm^{op}) mice results in male fertility defects. Biol Reprod 1996;55:310–7.
37. Hales DB. Leydig cell macrophage interactions: an overview. In: Payne AH, Hardy MP, Russell LD, eds. The Leydig cell. Vienna, IL: Cache River Press, 1996:451–65.

Part V

Toxicant Effects on Spermatogenesis

16

Estrogen Effects on Development and Function of the Testis

RICHARD M. SHARPE, JANE S. FISHER,
PHILIPPA T.K. SAUNDERS, GREGOR MAJDIC,
MIKE R. MILLAR, PRIYANKE PARTE,
JEFFREY B. KERR, AND KATIE J. TURNER

It is well established that exposure of the developing or adult male to even small amounts of exogenous estrogens can cause a range of adverse changes in reproductive development or function (1, 2). Such effects have been reported in laboratory and domestic animals as well as in man (1, 3). It has been generally presumed that these effects occur primarily as a consequence of the suppression of gonadotropin secretion (4), although there are suggestions in the literature that effects of estrogens within the testis or elsewhere in the male reproductive tract are possible (1, 2, 5).

In the past 3–4 years, two new developments have focused attention on the actions of estrogens in the male. First, it has been realized that some of the more ubiquitous man-made environmental chemicals are weakly estrogenic (1, 6), and it has been hypothesized that human exposure to such chemicals might have deleterious effects on male reproductive development, resulting in a decrease in sperm counts in adulthood (1, 4, 7). Second, the generation of estrogen receptor knockout mice (8) and the discovery of human males with naturally occurring mutations in the same gene (9) or in the gene that encodes aromatase (10, 11) have highlighted the concept that endogenous estrogen production in the male may be essential for normal reproductive function and fertility. The latter observations are particularly intriguing because the mechanistic basis for the impairment of reproductive function/fertility is not understood. It is our strong belief that aquisition of such understanding is a key issue, not only because it will identify the physiological role(s) of estrogens in the normal male, but also because it will provide the means for identifying precisely the mechanisms whereby exogenous estrogen exposure may perturb male reproductive development or function. This chapter reports on the different approaches that we are using to address this topic.

Effects of Neonatal Exposure to Estrogens on the Testis in Adulthood

To study the impact of neonatal estrogen exposure, male rat pups were injected on alternate days from either days 2 to 6 (3 injections) or days 2 to 12 (6 injections) after birth (day of birth = day 1) with either 20 μg diethylstilbestrol (DES; Sigma), 2 mg octylphenol (OP; Aldrich Chemical) or the injection vehicle (=20 μl corn oil). In these studies, OP was used as a representative "environmental estrogen," i.e., a weakly estrogenic man-made chemical (12). In an attempt to reduce between-litter errors, all males born on day 1 from several litters were "cross-fostered" randomly on the day of birth into all-male litters of 9–10 pups, and each of the three treatments was then represented in each litter. It was anticipated that one of the major effects of estrogen exposure during the neonatal period would be to suppress follicle-stimulating hormone (FSH) secretion and thus to reduce the rate of Sertoli cell proliferation that occurs in the rat to about day 15–16 (4, 13); this effect would be expected to result in smaller testes in adulthood but otherwise normal morphology (4, 7, 13, 14). Therefore, as a comparison, other male rats were injected on days 2 and 5 with 10 mg/kg of a potent long-acting gonadotropin-releasing hormone (GnRH) antagonist (Antarelix®, Europeptides, France) to suppress FSH secretion more or less completely during the neonatal period (14). Rats from all these treatment groups were allowed to grow to maturity and were eventually killed at age 90–95 days when testis size and morphology were evaluated. Representative animals from each treatment group were perfusion-fixed via the dorsal aorta (15) with cacodylate-buffered 2% glutaraldehyde.

Administration of 3 or 6 injections of DES resulted in reductions of 17% and 54%, respectively, in mean testis weight in adulthood when compared with littermate controls, while the much weaker environmental estrogen OP caused corresponding reductions of 7% and 12%, respectively (Table 16.1). Rats to which a GnRH antagonist had been administered neonatally had testis weights that averaged 49% lower than in respective control animals (Table 16.1). One unexpected finding in the study in which DES and OP were administered was that control rats that had received 6 injections of vehicle neonatally had significantly lower testis weights than did those which received only 3 injections (Table 16.1). This was not considered to be a chance effect because it occurred consistently in several litters and the animals had testis weights below the normal control range for our rat colony. We consider that the most likely explanation for this finding is that DES excreted by the DES-treated pups in each litter had been ingested by the mother during grooming of the pups and had then been recycled back to all of the pups (including control and OP-treated animals) via the breast milk and that this had exerted a small background effect on testis size. If this interpretation is correct, it is noteworthy that the much weaker estrogen OP was still able to suppress testis weight slightly, but significantly, in such animals despite the background presence of the potent DES (Table 16.1).

TABLE 16.1. Effect of neonatal injection of diethylstilbestrol (DES) or octylphenol (OP) on testicular weight in adulthood (90–95 days) in rats in comparison with animals in which follicle-stimulating hormone (FSH) was suppressed neonatally by the administration of a long-acting gonadotropin-releasing hormone (GnRH) antagonist.

Neonatal treatment (days)	Treatment	No. of animals	Testis weight (mg)	Testis weight as % control (mean ± SD)
2–6[a]	Vehicle	10	1921 ± 79	(100)
	DES (20 μg)	9	1588 ± 115***	82.6
	OP (2 mg)	10	1774 ± 91**	92.3
2–12[b]	Vehicle	10	1614 ± 109[c]	(100)
	DES (20 μg)	8	751 ± 240***[c]	46.5[d]
	OP (2 mg)	13	1419 ± 154**[c]	87.9
2–5[e]	Vehicle	6	1978 ± 163	(100)
	GnRH antagonist (10 mg/kg)	8	1012 ± 99***	51.0

[a]Injected on days 2, 4, and 6.
[b]Injected on days 2, 4, 6, 8, 10, and 12.
[c]$p < .001$, in comparison with respective group treated for 2–6 days only.
[d]Excludes animals with necrotic testes (see text).
[e]Injected on days 2 and 5.
**$p < .01$, ***$p < .001$, in comparison with respective vehicle-treated (control) group.

In all except one of the groups of animals, testicular morphology was grossly normal with complete spermatogenesis in all seminiferous tubules and with seminiferous tubule diameters within the control range (Fig. 16.1). The exception was animals that had been administered 6 injections of DES, as these presented with one of three different morphological variations. Approximately 36% of this group exhibited grossly normal spermatogenesis, although with somewhat reduced seminiferous tubule diameter, 36% exhibited essentially Sertoli-cell-only syndrome (these had the lowest testis weights), and in the remainder (28%) either one or both testes were completely necrotic (Fig. 16.1). The latter testes had developed adhesions to the surrounding membranes and were firmly "stuck" in the inguinal canal. In contrast to the heterogeneous morphology of the latter treatment group, animals treated neonatally with a GnRH antagonist exhibited grossly normal spermatogenesis and seminiferous tubule diameter (not shown).

Using the GnRH antagonist-treated rats as a reference for the consequences of suppressing FSH (and luteinizing hormone, LH) secretion during neonatal life, it is clear that the reduction in testis size in the OP-treated and some of the DES-treated animals is probably explained simply by suppression of FSH caused by administration of "estrogens," resulting in fewer Sertoli cells (4, 13). This conclusion would be consistent with the absence of any discernible change in testicular morphology. However, suppression of FSH (or LH) in rats to which had been administered 6 injections of DES cannot explain the 64% occurrence of Sertoli-cell-only and necrotic testes in this treatment group, because no such changes were found in the GnRH anagonist-treated rats in which the most complete

170   R.M. Sharpe et al.

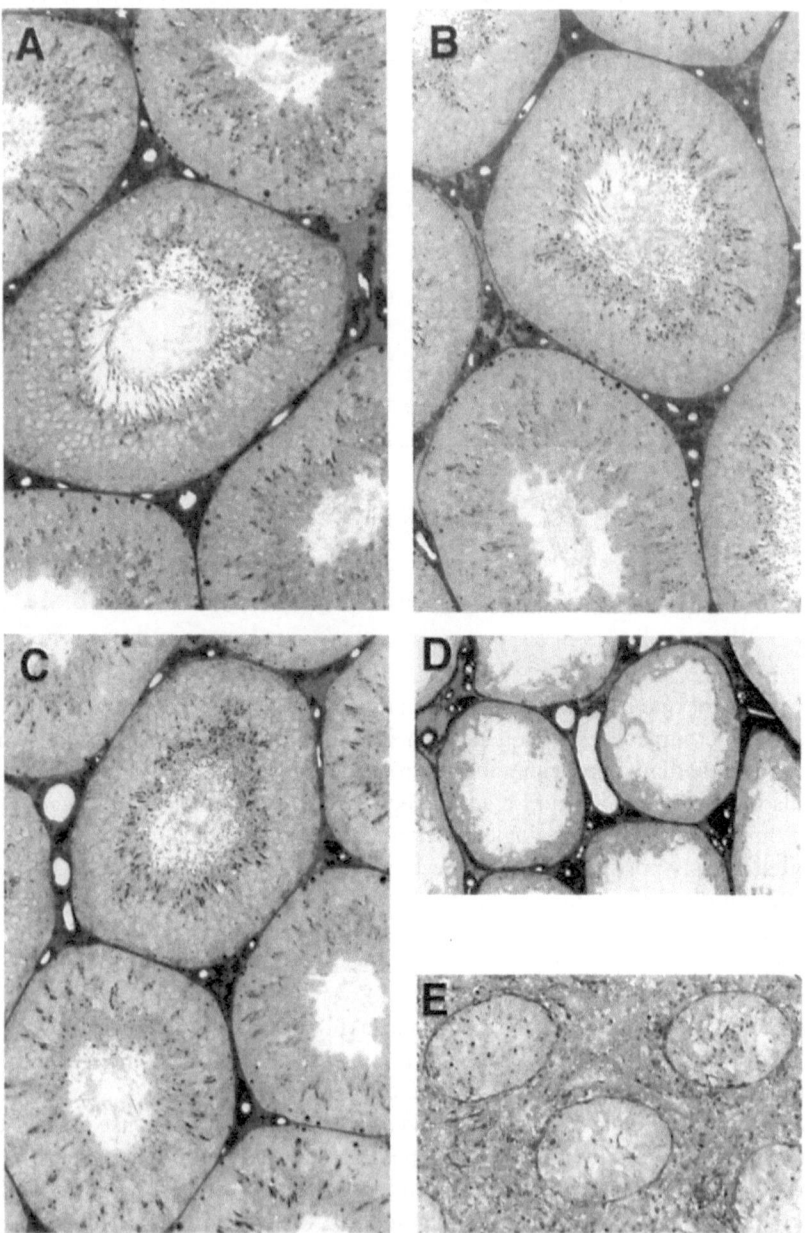

FIGURE 16.1. Representative testicular morphology in adult rats treated neonatally from days 2 to 12 with either vehicle (*A*), octylphenol (*B*), or diethylstilbestrol (*C, D, E*). Note that morphology is grossly normal in octylphenol-treated rats and in approximately one-third of diethylstilbestrol-treated rats (*C*), although in the latter seminiferous tubule diameter was slightly reduced. However, nearly two thirds of diethylstilbestrol-treated rats exhibited either Sertoli-cell-only testes (D) or had necrotic testes (E). ×200.

suppression of FSH and LH secretion occurred (14). It is therefore concluded tentatively that the DES-induced changes resulting in Sertoli-cell-only or necrotic testes reflect additional direct effects of DES within the testis, possibly on the Sertoli cells or the germ cells (see following). In this context, it is important to remember that these effects were not observed in animals treated with DES on postnatal days 2–6 but only in animals administered DES on days 2–12. This presumably means either that some estrogen-modulable event occurs specifically during days 6–12 or shortly thereafter, or that the longer period of DES exposure has prevented a compensatory change (e.g., in Sertoli cell development) during days 6–12 that was possible in rats affected by DES during days 2–6.

The problem with the studies just described and those in the literature reporting various adverse effects of estrogen administration to animals is that they must be interpreted without good information as to where and when estrogens act physiologically in the male reproductive system. We therefore decided that a valuable approach would be to initiate research to identify, first, where estrogens act in the male by locating where estrogen receptors were expressed and, second, to identify what molecular and biochemical processes were regulated by estrogens at these sites.

Sites of Estrogen Receptor Expression in the Male Reproductive System

Using immunohistochemistry and optimized methods, receptors for estrogen receptor-α (ERα) were localized using a commercially available antiserum (Novocastra) that is thought unlikely to cross-react with the recently reported ERβ receptor (16,17). In both the rat and marmoset monkey, an essentially similar pattern of expression of ERα was observed (18). Within the neonatal testis, the fetal Leydig cells exhibited high immunoexpression of ERα, whereas in the pubertal/adult testis immuonexpression of ERα in the adult generation of Leydig cells was much less intense (18). At no age was there any detectable expression of ERα in Sertoli cells, germ cells, macrophages, or the vasculature (18). In the rat, but not in the marmoset monkey, ERα was expressed in the epithelial cells of the rete testis (not shown), but in both animals the most pronounced expression of ERα occurred in the epithelial cells of the efferent ducts (Fig. 16.2). In neither species was major expression of ERα observed in the epididymis, although in the rat some immunostaining was detected in the initial segment around the time of puberty (18).

Clearly, comparable studies with antibodies specific for ERβ are necessary before the full range of sites of estrogen action in the male reproductive system can be identified, but, on the basis of our findings with ERα, some sites of estrogen action can be unequivocally stated. First, it is clear that Leydig cells, and possibly their precursors, are a site of estrogen action, a conclusion that would fit with various pieces of biochemical evidence (19, 20). Second, the efferent ducts are clearly the major site of estrogen action, judging from the intense immunoex-

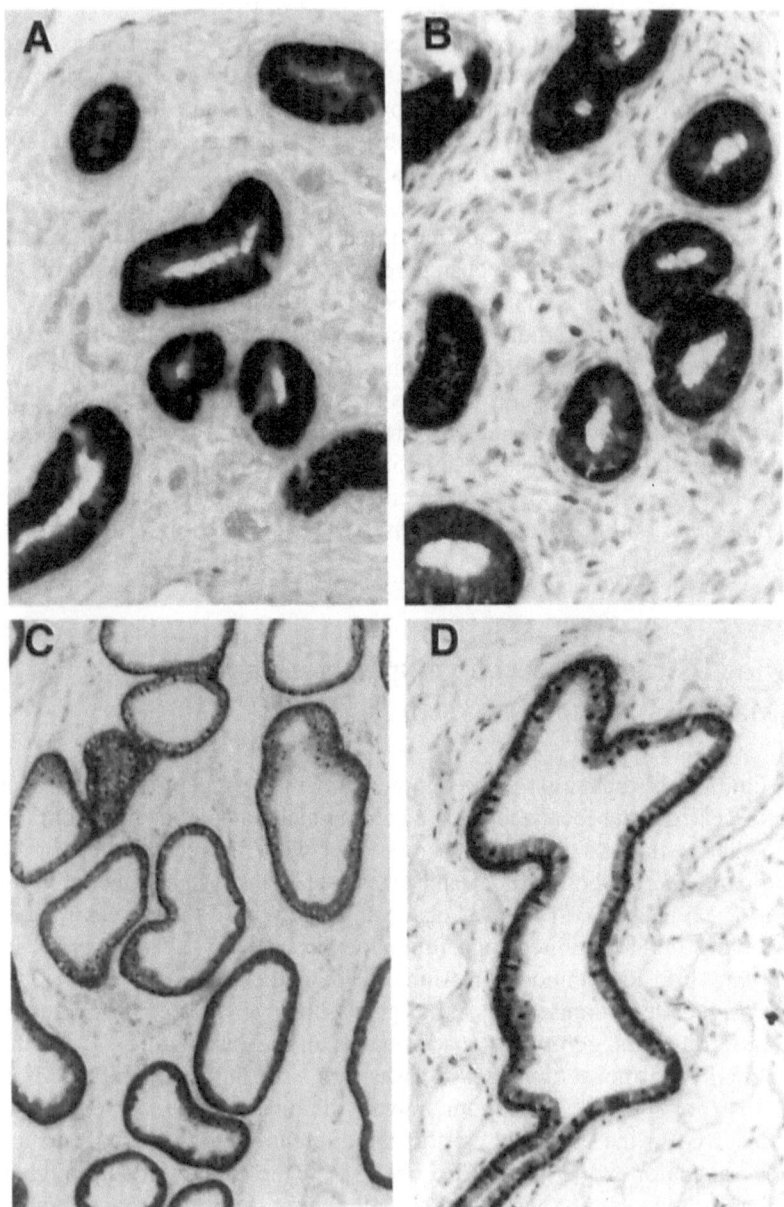

FIGURE 16.2. Immunoexpression of estrogen receptor-α in the epithelium of the efferent ducts of rats at 8 (*A*), 15 (*B*), or 90 days of age (*C, D*). *A, B, D,* ×400; *C,* ×200.

pression of ERα. As the one known function of the efferent ducts is to reabsorb fluid from around the spermatozoa leaving the testis (21), it is reasonable to propose that estrogens may somehow be involved in the modulation of this process. In this respect, there are some interesting recent findings to support this line of thinking. First, it has been shown that the male ERα-knockout (ERKO) mouse exhibits distension of the rete testis and of the lumens of seminiferous tubules (8), consistent with impairment of fluid resorption from the efferent ducts. Second, it has been shown that sperm express aromatase as they leave the testis and traverse the efferent ducts (22, 23), raising the possibility that it is the sperm themselves that regulate fluid resorption in the efferent ducts via the local genera-tion of estradiol from the plentiful supply of testosterone in the surrounding fluid.

Effect of Administration of a Specific Aromatase Inhibitor to Adult Male Rats

In conjunction with the studies to identify sites of expression of ERα, we have also initiated studies to examine the consequences of selectively suppressing endogenous estrogen production using a new and highly specific aromatase inhib-itor, anastrozole (marketed as Arimidex®; Zeneca plc, UK). This compound was added to the drinking water of adult male rats at a concentration of 100 mg/l for 7 weeks, at which time the rats were either killed or mated. In addition, blood samples were obtained at intervals during the 7-week treatment from the tail vein under light halothane anesthesia. The preliminary findings from these studies are as follows. First, none of the anastrozole-treated animals (n = 7) mated with female rats during cohabitation for 1 week (off treatment during this time), as assessed by the presence of ejaculatory plugs, whereas 8 of 9 control males successfully mated. Second, after 1 week of anastrozole treatment, the blood levels of FSH had increased by 1.6 ± 0.4 ng/ml (mean ± SD; $p < .001$) compared with pretreatment levels, whereas no significant change (-0.1 ± 0.2 ng/ml) occurred in controls. Third, examination of the efferent ducts from perfusion-fixed animals showed consistent differences between control and anastrozole-treated rats (Fig. 16.3), with ducts from the latter exhibiting larger lumens and a more flattened epithelium (presumably caused by to the distension of the lumen). Fourth, examination of the testes revealed grossly normal spermatogenesis al-though the lumens of seminiferous tubules appeared to be somewhat larger than in the controls, and this was reflected by a small but significant increase in testicular weight (not shown); Leydig cell numbers were normal or perhaps slightly reduced as assessed subjectively (not shown).

Although the degree of suppression of aromatase was not determined in these studies, the absence of mating (which is estrogen dependent in the rat) (24) and the increase in blood FSH levels (10, 11) are consistent with effective suppression of aromatase in anastrozole-treated animals. The distension of the lumen of the efferent ducts observed in anastrozole-treated rats is also consistent with lowered

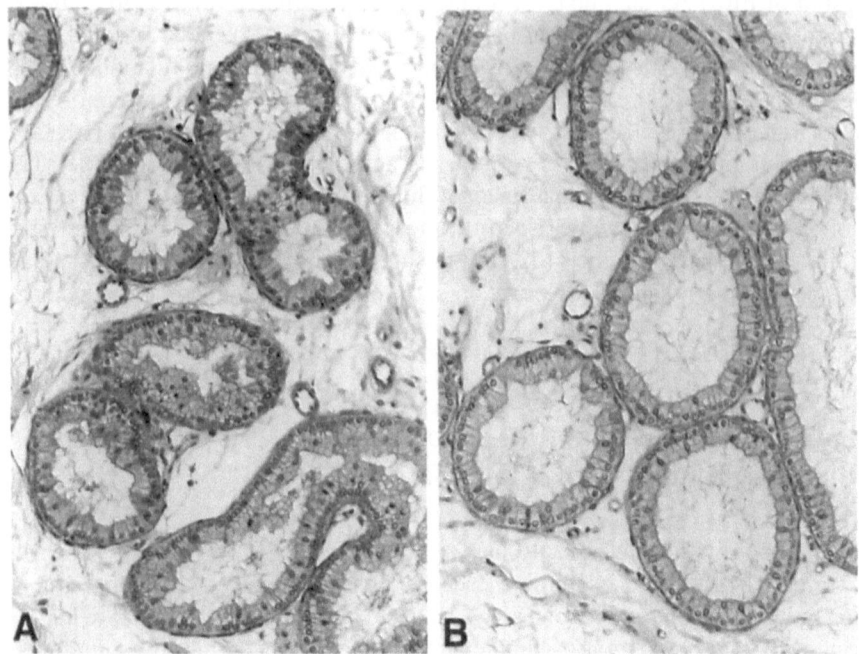

FIGURE 16.3. Cross-sectional appearance of the efferent ducts in a control rat (*A*) and in a rat treated for 7 weeks previously with the aromatase inhibitor anastrozole (*B*). ×200.

estrogen production/action based on findings in ERKO mice (8) and the intense expression of ERα in the epithelium of these ducts, as described here. We therefore concluded from these preliminary studies that the administration of anastrozole to rats should provide an effective tool with which to explore the role(s) of estrogens in the male.

Summary

The precise role(s) of estrogens in the male reproductive system is largely unknown and thus our ability to understand when and how estrogens can interfere with normal reproductive development and function in the male remains extremely superficial. We hope this short chapter has outlined some of the approaches that can be used to fill the gaps in our understanding of estrogen action in the male. These and other studies have highlighted the efferent ducts as a major site of estrogen action, and it seems reasonable that estrogens, probably deriving from spermatozoa, are involved in modulating fluid resorption at this site. However, it is puzzling that, during neonatal and even fetal life, receptors for ERα are expressed at high levels in the efferent ducts and mesonephros (Fig. 16.2) (18). These findings raise the possibility that estrogens exert actions at this site long before spermatozoa are produced or there is need to reabsorb fluid; precisely what these actions are remain to be determined. Within the testis, ERα are expressed in Leydig cells and possibly in their precursors, and this would be consistent with evidence for a negative role of estrogens in the regulation of steroidogenic enzyme activity (20) and Leydig cell development (13, 19). In view of the latter, hyperplasia of the Leydig cells was anticipated in rats treated for 7 weeks with anastrozole, but none was observed; indeed, the impression gained was that there might be fewer Leydig cells in these animals, although this clearly needs rigorous and objective evaluation.

Judged on the sites of expression of ERα and the effects of suppressing FSH/LH secretion during neonatal life, it is difficult to explain the induction of Sertoli-cell-only syndrome in a third of animals treated on neonatal days 2–12 with DES. However, as the recently described second estrogen receptor, ERβ, is reportedly expressed in the testis (16) and a variety of in vitro studies have reported actions of estrogens on Sertoli cells from immature rats (25) and on gonocytes from fetal rats (5), it is possible that the effects of DES in this situation might be mediated via this receptor rather than via ERα. Immunolocalization studies with antibodies specific to ERβ are necessary to confirm whether this is a possibility.

Acknowledgments. We are grateful to Jim McDonald, Denis Doogan, and Sheila McPherson for their skilled technical help and to Dr. Mike Dukes (Zeneca plc) and Dr. R. Deghenghi (Europeptides) for gifts of Anastrozole and Antarelix, respectively. K.J.T. was supported by grant number PL950314 from the Commission of the European Communities, Research on Occupational and Environmental Health (BIOMED 2).

References

1. Toppari J, Larsen JC, Christiansen P, Giwercman A, Grandjean P, Guillette LJ Jr, et al. Male reproductive health and environmental xenooestrogens. Environ Health Perspect 1996;104((suppl 4):741–803.
2. Arai Y, Mori T, Suzuki Y, Bern HA. Long-term effects of perinatal exposure to sex steroids and diethylstilbestrol on the reproductive sytem of male mammals. Int Rev Cytol 1983;84:235–68.
3. Sharpe RM. Could environmental oestrogenic chemicals be responsible for some disorders of human male reproductive development? Curr Opin Urol 1994;4:295–301.
4. Sharpe RM. Declining sperm counts in men—is there an endocrine cause? J Endocrinol 1993;136:357–60.
5. Li H, Papadopoulos V, Vidic B, Dym M, Culty M. Regulation of rat testis gonocyte proliferation by platelet-derived growth factor and estradiol: identification of signalling mechanisms involved. Endocrinology 1997;138:1289–98.

6. Turner KJ, Sharpe RM. Environmental oestrogens—present understanding. Rev Reprod 1997;2:69–73.
7. Sharpe RM, Skakkebaek NE. Are oestrogens involved in falling sperm counts and disorders of the male reproductive tract? Lancet 1993;341:1392–5.
8. Eddy EM, Washburn TF, Bunch DO, Goulding EH, Gladen BC, Lubahn DB, et al. Targeted disruption of the estrogen receptor gene in male mice causes alteration of spermatogenesis and infertility. Endocrinology 1996;137:4796–805.
9. Smith EP, Boyd J, Frank GR, Takahashi H, Cohen RM, Specker B, et al. Estrogen resistance caused by a mutation in the estrogen receptor gene in man. N Engl J Med 1994;331:1056–61.
10. Morishima A, Grumbach MM, Simpson ER, Fisher C, Qin K. Aromatase deficiency in male and female siblings caused by a novel mutation and the physiological role of oestrogens. J Clin Endocrinol Metab 1995;80:3689–98.
11. Qin K, Fisher CR, Simpson ER, Serpente S, Faustini-Fustini M, Carani C. Oligospermia and persistent linear growth of a male subject caused by a mis-sense mutation in the gene encoding aromatase. In: Program & Abstracts of 10th International Congress of Endocrinology, San Francisco, 1996 (Abstract P2–733).
12. White R, Jobling S, Hoare SA, Sumpter JP, Parker MG. Environmentally persistent alkylphenolic compounds are estrogenic. Endocrinology 1994;135:175–82.
13. Sharpe RM. Regulation of spermatogenesis. In: Knobil E, Neill JD, eds. The physiology of reproduction, 2nd Ed. New York: Raven Press, 1994:1363–464.
14. Kolho K-L, Huhtaniemi I. Suppression of pituitary-testis function in rats treated neonatally with a gonadotrophin-releasing hormone agonist and antagonist: acute and long-term effects. J Endocrinol 1989;123:83–91.
15. Kerr JB, Millar M, Maddocks S, Sharpe RM. Stage-dependent changes in spermatogenesis and Sertoli cells in relation to the onset of spermatogenic failure following withdrawal and restoration of testosterone. Anat Rec 1994;235:547–59.
16. Kuiper GGJM, Enmark E, Pelto-Huikko M, Nilsson S, Gustafsson J-A. Cloning of a novel oestrogen receptor expressed in rat prostate and ovary. Proc Natl Acad Sci USA 1996;93:5925–30.
17. Kuiper GGJM, Carlsson B, Grandien K, Enmark E, Haggblad J, Nilsson S, et al. Comparison of the ligand binding specificity and transcript tissue distribution of estrogen receptors α and β. Endocrinology 1997;138:863–70.
18. Fisher JS, Millar MR, Majdic G, Saunders PTK, Fraser HM, Sharpe RM. Immunolocalisation of oestrogen receptor-α (ERα) within the testis and excurrent ducts of the rat and marmoset monkey from perinatal life to adulthood. J Endocrinol 1997;153: 485–95.
19. Abney TO, Myers RB. 17β-Oestradiol inhibition of Leydig cell regeneration in the ethane dimethylsulfonate-treated mature rat. J Androl 1991;12:295–304.
20. Cigorraga SB, Sorrell S, Bator J, Catt KJ, Dufau ML. Oestrogen-dependence of a gonadotropin-induced steroidogenic lesion in rat testicular Leydig cells. J Clin Invest 1980;65:699–705.
21. Ilio KY, Hess RA. Structure and function of the ductuli efferentes. Microsc Res Tech 1994;29:432–67.
22. Nitta H, Bunick D, Hess RA, Janulis L, Newton SC, Millette CF, et al. Germ cells of the mouse testis express P450 aromatase. Endocrinology 1993;132:1396–401.
23. Janulis L, Hess RA, Bunick D, Nitta H, Janssen S, Osawa Y, et al. Mouse epididymal sperm contain active P450 aromatase which decreases as sperm traverse the epididymis. J Androl 1996;17:111–6.

24. Meisel RL, Sachs BD. The physiology of male sexual behavior. In: Knobil E, Neill JD, eds. The physiology of reproduction, 2nd Ed., Vol. 2. New York: Raven Press, 1994:3–106.
25. Panno ML, Sisci D, Salerno M, Lanzino M, Mauro L, Morrone EG, et al. Effect of triiodothyronine administration on estrogen receptor contents in peripubertal Sertoli cells. Eur J Endocrinol 1996;134:633–8.

17

Developmental Effects and Molecular Mechanisms of Environmental Antiandrogens

WILLIAM R. KELCE AND ELIZABETH M. WILSON

Industrial chemicals and environmental pollutants can disrupt reproductive development in wildlife and humans by altering the synthesis, transport, action, or elimination of gonadal steroid hormones. Steroid hormones control fundamental events in embryonic development and sex differentiation by binding to their cognate nuclear receptors, which act as steroid-inducible transcription factors to activate or repress transcription of target genes. The consequences of disrupting these events can be especially profound during embryonic development because the role of steroid hormones is crucial in controlling transient and irreversible developmental processes. Recent studies suggest that certain industrial pollutants and environmental pesticides have the potential to alter male sex development and reproductive processes in wildlife and human populations by acting as environmental antiandrogens (1–3). In some cases, laboratory studies have confirmed abnormalities of reproductive development observed in the field and have provided mechanisms to explain the disruptive effects of these environmental chemicals.

Reports of the increasing incidence of developmental reproductive tract abnormalities (e.g., hypospadias) in the human male population and decreased adult sperm counts in some parts of the world, together with the recent identification of major pesticides that have antiandrogenic activity, necessitate a closer look at the molecular and functional implications of antiandrogens in the environment. To this end, we briefly review proposed mechanisms by which androgen receptor (AR) agonists and antagonists induce or inhibit transcription of AR-dependent genes. We discuss the role of androgens and AR in sex differentiation and development and the developmental effects and molecular mechanisms of vinclozolin and p,p'-DDE, recently discovered environmental chemicals with antiandrogenic activity. As opinions vary regarding the best approach to screen for endocrine disruptors, that is, primarily in vitro versus in vivo assays, we propose a comprehensive investigational strategy to identify antiandrogenic chemicals or metabolites and provide data concerning the mechanism responsible for the phenotypic developmental effects. We focus on effects occurring during development, because many of these effects are permanent and are likely to influence risk assessment decisions. The reader is referred to literature citations contained within several recent reviews for discussions of general mechanisms of environmental endocrine disruptors (4–6), effects of environmental endocrine disruptors in wildlife (7, 8), effects of environmental estrogens in the male (9), clinical implications of these chemicals in humans (6, 10), and proposed research needs for risk assessment (11).

Mechanisms of Androgen/Antiandrogen Action

The androgen testosterone produced by the Leydig cells of the testis dissociates from carrier proteins in human plasma and diffuses into cells where it binds AR. In target tissues such as prostate and epididymis, testosterone is reduced to the more potent androgen, 5α-dihydrotestosterone (DHT). The enhanced androgenic activity of DHT results in part from its slower rate of dissociation from AR (12). High-affinity androgen binding induces a conformational change in the AR that is requisite for stabilization against proteolytic degradation, a protective mechanism intrinsic to and facilitated by the AR NH_2-terminal domain (12). Agonist-bound AR undergoes conformational changes likely resulting in the loss of heat-shock proteins and is imported into nuclei where it dimerizes (13) and binds to androgen response element DNA regulatory sequences within intron regions or flanking androgen-responsive genes, resulting in transcriptional activation of those genes (14, 15). During sex differentiation and development, these gene products carry out specific androgen-dependent cell functions critical to the sex differentiation of developing tissues.

Antihormone binding is thought to induce a steroid receptor conformation that differs from that imposed by agonist binding, thereby altering its ability to activate transcription (16–18). Proposed models of antihormone action include two mechanisms of transcriptional inhibition that reflect the ability of receptors to bind DNA. Type I antagonists bind receptors but prevent DNA binding and thus transcriptional activation, whereas type II antagonists promote DNA binding but nevertheless fail to initiate transcription (19). Environmental antiandrogens identified to date exhibit low to moderate affinity for AR and act as type I antagonists by preventing AR binding to androgen response element DNA (1, 20). We have previously suggested that binding of two different ligands in the same receptor dimer (mixed-ligand dimer) is required for AR antagonism (20), but the specific mechanisms responsible for inhibition of AR–DNA binding remain to be established. They could include increased AR degradation via an inappropriate receptor conformation or the inability of rapidly dissociating ligands to stabilize AR (12, 21). Alternatively, the AR dimerization interface for agonist- and antagonist-bound AR may be incompatible, and no mixed-ligand dimers form. Finally, mixed-ligand AR dimers may fail to bind DNA because of inappropriate dimer conformation (22) or inability to release receptor-associated proteins requisite for subsequent DNA binding (23, 24).

Sex Differentiation: Role of Androgens and the Androgen Receptor

Genetic alterations in sex differentiation and development are not lethal and therefore provide information regarding the role of sex steroids and their receptors in mammalian reproductive development (25, 26). Genetic sex is determined at fertilization by the presence of the Y chromosome that directs differentiation of the indifferent gonads into testes. Before this time, the embryo can potentially develop into the male or female phenotype. Testicular testosterone secretion induces differentiation of the Wolffian duct into epididymis, vas deferens, and seminal vesicles. The testosterone metabolite, 5α-dihydrotestosterone (DHT), induces development of the prostate and male external genitalia. This apparent differential activity of androgens most likely occurs at the level of androgen metabolism (27) as both testosterone and DHT bind the same receptor (28). In the absence of testosterone, the female phenotype is expressed independent of the presence of an ovary.

Immunostaining and reverse transcriptase-polymerase chain reaction analysis indicates AR is present in most tissues; however, the highest levels of expression are in the male reproductive tract (29). A single AR gene located on the long arm of the X chromosome at Xq11-12 (30–32) encodes a single AR protein of apparent 110–114 kDa, comprising amino acids 910–919 (33–35). The variability in AR length results from polymorphic glutamine (CAG) and glycine (GGC) repeats in the NH_2-terminal domain (34–40). AR mediates sex differentiation in the developing fetus as well as androgen action in postembryonic life, as immunoblots prepared from urogenital tract tissues of gestational day 17 male and female rats recognize the same 110-kDa protein characteristic of AR from adult tissues (41). In addition, the steroid-binding characteristics of AR isolated from the embryonic urogenital tract are identical to those determined from mature adult reproductive tract tissues (41). The greater sensitivity of the developing male fetus to the antiandrogenic effects of environmental chemicals may result from reduced levels of competing endogenous androgens in the developing fetal male compared to the adult.

Basic mechanisms of sex differentiation are similar in mammals, so chemicals that affect reproductive development in rodents and other mammals should be considered potential human reproductive toxicants. Species differences in health effects of these chemicals undoubtedly depend on their ability to metabolize or distribute these chemicals; this is even true among individuals of different ages within a species. Chemical exposure during sex differentiation is of particular concern for several reasons. First, development of the reproductive system often is sensitive to low-dose chemical effects. Second, while chemical exposure may be transient, the effects can be irreversible. Third, functional alterations often are not discovered until well after exposure (i.e., at puberty or later in life), leading to underestimates of chemical-induced effects on reproductive development. Finally, developmental abnormalities cannot be predicted from chemical exposures in adult animals because adults are fully differentiated and can therefore tolerate a greater chemical insult. Developmental reproductive toxicity data, then, are critical for the assessment of noncancer health effects of endocrine-disrupting chemicals.

Environmental Antiandrogens: Developmental Effects and Molecular Mechanisms

Vinclozolin

The fungicide vinclozolin (Fig. 17.1) has antiandrogen activity and alters sex differentiation in male rats by inhibition of AR-mediated gene activation (2, 20). In male rat offspring, perinatal exposure to vinclozolin causes hypospadias, ectopic testes, vaginal pouch formation, agenesis of the ventral prostate, and nipple retention; females are phenotypically normal (3). Exposure to 50 mg vinclozolin/kg body weight per day from gestational day 14 to postnatal day 3 induces infertility and reduced ejaculated sperm counts in adult male offspring as a result of severe hypospadias. Concentrations as low as 12 mg kg^{-1} day^{-1} permanently reduce ventral prostate weight following developmental exposure. In contrast, fertility is unaffected in adult male rats even following prolonged high-dose exposure (100 mg kg^{-1} day^{-1} for 25 weeks) (42). It follows, then, that the developing fetus is particularly sensitive to endocrine disruptors such as vinclozolin, which can produce malformations at dosages that have little or no reproductive effect in adult males. The extent to which the human population is exposed to vinclozolin or its active metabolites either directly or via the food chain remains to be established.

The molecular mechanisms responsible for the antiandrogenic effects of vinclozolin have been elucidated (2, 20). Vinclozolin is hydrolyzed to two ring-opened metabolites, M1 and M2, which compete for AR androgen binding, induce AR nuclear import, and inhibit DHT-induced transcriptional activation by blocking AR binding to androgen response element DNA (20). The parent chemical vinclozolin is a poor inhibitor of AR–androgen binding and subsequent transactivation, suggesting that vinclozolin developmental toxicity is likely mediated via the formation of the active metabolites M1 and M2 (2). The enhanced activity of the vinclozolin metabolites, compared to vinclozolin itself, was perhaps predictable based on the structural similarity of these metabolites with the potent antiandrogen hydroxyflutamide (Fig. 17.1).

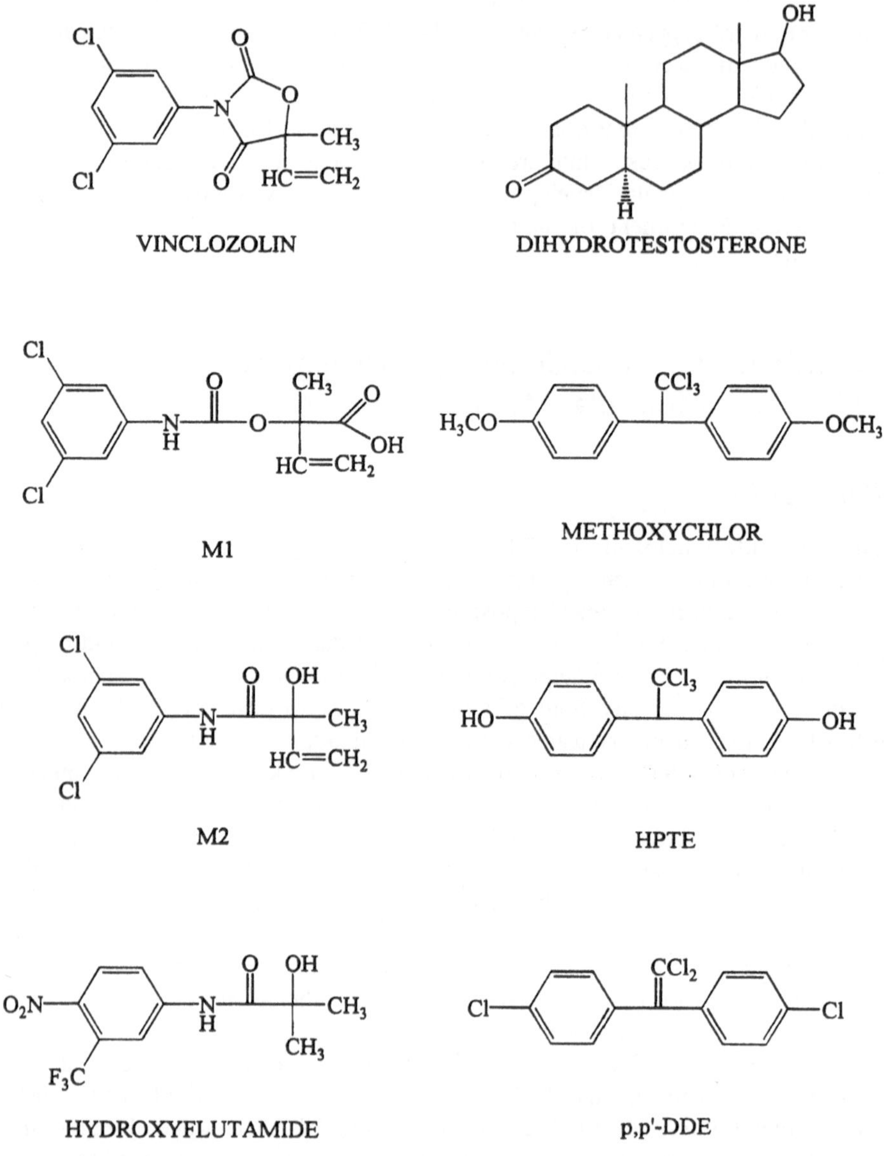

VINCLOZOLIN   DIHYDROTESTOSTERONE

M1   METHOXYCHLOR

M2   HPTE

HYDROXYFLUTAMIDE   p,p'-DDE

FIGURE 17.1. Structural formulas of vinclozolin, vinclozolin metabolites M1 and M2, 5α-dihydrotestosterone, *p,p'*-DDE, hydroxyflutamide, methoxychlor, and the methoxychlor metabolite, 2,2-bis-*p*-hydroxyphenol-1,1,1-trichloroethane (HPTE).

p,p'-DDE

Another environmental chemical recently identified as a potent antiandrogen is p,p'-DDE, a p,p'-DDT metabolite that bioaccumulates in the environment (1). p,p'-DDE is detectable in food (43, 44) and in human body fat (45, 46), has a half-life in the body and the environment of more than 65 years, and constitutes 50%–80% of the total DDT-derived residues in human breast milk mobilized to the suckling infant (47–49). p,p'-DDE is often confused in the literature as being an environmental estrogen, but it does not bind the estrogen receptor (1). However, o,p'-DDT, a minor DDT isomer, binds the estrogen receptor and has estrogenic activity (50). When administered to pregnant rats from gestational day 14 to 18, p,p'-DDE (100 mg kg^{-1} day^{-1}) reduces anogenital distance and causes retention of thoracic nipples in male progeny (1), both being indicative of prenatal antiandrogen exposure (51). p,p'-DDE binds AR in vitro with moderate affinity and inhibits DHT-induced transcriptional activation with a potency similar to that of the antiandrogenic drug hydroxyflutamide (1). Similar mechanistic observations have been made with the structurally related pesticide, methoxychlor, and its primary o-demethylated metabolite, 2,2-bis-p-hydroxyphenol-1,1,1-trichloroethane (unpublished observations).

Although the use of DDT for agricultural purposes was banned in the United States in 1973, its metabolites persist in the environment, and DDT and DDE continue to be used as pesticides in other countries. Recent evidence of a global distillation effect for some organochlorine pollutants indicates migration through the atmosphere from warmer to colder latitudes (52). Median levels of p,p'-DDE measured in serum (12.6 ppb) and placental tissues (6.8 ppb) from women in the United States within the past 10 years are less than those required to inhibit androgen action in vitro. However, maximum levels in certain populations exceed these concentrations (e.g., 180 ppb in serum and 74 ppb in placenta) (49), presumably as the result of continuous low-level exposure and slow metabolic clearance. Bouwman et al. (53) reported median DDT/DDE serum levels of 140.9 and 475 ppb in breast milk from individuals living in South African dwellings treated with DDT for mosquito control against the spread of malaria, leading them to postulate a significant health risk to infants (54). In the mid-1960s, when DDT was in use in the United States, high concentrations of p,p'-DDE were measured in tissues from stillborn infants in Atlanta, Georgia (650 ppb in brain; 850 ppb in lungs; 2740 ppb in heart; 980 ppb in liver; 3570 ppb in kidney; 860 ppb in spleen) (55). Although the in vivo cellular concentration of p,p'-DDE is not known, reports suggest that human p,p'-DDE levels can exceed those that inhibit human AR transcriptional activation in vitro.

Birth defects in humans that could result from inhibition of AR action include alterations in the development of the male external genitalia (56). Environmental chemicals with antiandrogen activity may contribute to the increasing incidence of isolated hypospadias in the human population, as this anomaly is not frequently associated with mutations in the AR coding sequence (57, 58), with steroid 5α-reductase (type 2) deficiency (59), or with decreased levels of AR (60). Hypospadias has been linked to fetal exposure to estrogenic chemicals in the first trimester (61); however, the external genitalia of the human male fetus may lack estrogen receptor (62). As estrogens bind AR with moderate affinity (28), the increased incidence of hypospadias following developmental exposure to estrogenic chemicals may be mediated at the level of the AR. Environmental antiandrogens could induce male pseudohermaphroditism (incomplete masculinization of the male fetus) to differing degrees depending on time of exposure and chemical potency (63).

Investigational Strategies to Screen for Environmental Endocrine Disruptors

Multigenerational reproductive toxicology test protocols do not currently require endocrine data or even bioassay measurements for hormone activity (64, 65). Laboratories at the U.S. Environmental Protection Agency have developed multigenerational test guidelines that include measures of endocrine function such as pubertal landmarks, estrous cyclicity, and reproductive organ weights in an alternate reproduction test protocol (66, 67). Because these tests are labor intensive, require numerous animals, take more than a year to complete, and are expensive, shorter-term test protocols are sought to identify endocrine-disrupting chemicals and to elucidate their mechanisms of endocrine toxicity. It is doubtful that single in vivo or in vitro tests will adequately assess endocrine-disrupting activity. For example, screening chemicals in vitro for estrogenicity may fail to identify compounds that act through other steroid hormone receptors, alter steroid hormone biosynthesis, transport, or degradation, or require metabolic activation. Such in vitro estrogenicity tests alone would fail to predict the antiandrogenic in vivo effects induced by vinclozolin or p,p'-DDE.

We propose combined in vivo and in vitro test strategies to screen chemicals for reproductive toxicity. In vivo tests identify chemicals with endocrine-disrupting activity, whereas in vitro tests reveal the effective chemical or metabolite and provide information regarding the biochemical mechanism. Given this information, human susceptibility and risk assessment issues can be addressed subsequently. Dosing newborn animals with toxicants has been a successful strategy in detecting estrogenicity and antiandrogenicity. The age and weight at puberty, weights of the reproductive organs, and serum hormone levels are measured from day 22 to 50 in male and female rats that were treated since birth. This approach has revealed the toxic effects of chlordecone, methoxychlor, vinclozolin, and DDE.

The proposed strategy combines in vivo studies to identify endocrine disrupting chemicals with in vitro studies to characterize the molecular mechanism of action. Using this protocol, vinclozolin administered for 30 days starting at weaning delayed puberty, reduced the weights of the sex accessory glands, and increased serum testosterone and luteinizing hormone (LH) levels consistent with the endocrine profile of an antiandrogen. In a developmental study, vinclozolin caused alterations in male rat sex differentiation such as reduced anogenital distance, cleft phallus, hypospadias, ectopic testes, and retained thoracic nipples in male offspring (3), all indicative of antiandrogen activity. In vivo experiments such as these helped to identify vinclozolin as an antiandrogen endocrine disruptor.

Subsequent in vitro studies demonstrated that although the parent chemical, vinclozolin, bound AR weakly, its two primary hydrolysis products were more potent androgen antagonists (2). Maternal serum concentrations of these metabolites were at levels near the K_i for inhibition of androgen binding, suggesting that the developmental toxicity of vinclozolin is mediated by its hydrolysis products, M1 and M2 (2). Molecular studies determined that the mechanism of inhibition of androgen-induced transcriptional activation was inhibition of AR–DNA binding (20). Within a relatively short period and with a limited number of animals, the antiandrogenic activity of vinclozolin was identified, its adverse developmental effects characterized, and the biochemical and molecular mechanism established. Thus, combining in vivo and in vitro test strategies is effective in the identification and characterization of environmental endocrine disruptors.

The foregoing testing strategy was used to identify more than 25 environmental chemicals predicted to have an affinity for AR. Those structures shown biochemically to interact with AR were introduced into a computer model together with known androgen agonists and antagonists to develop a three-dimensional quantitative structure–activity relationship (3D-QSAR) paradigm that predicts AR binding affinity solely from chemical structure, taking into account steric and electrostatic properties (68). The model is being used to search structural databases for potential androgen agonists and antagonists. Chemicals identified using the computer are subsequently examined for AR-binding activity and induction or inhibition, respectively, of androgen-dependent transcriptional activity. Chemicals that alter androgen action in vitro are further tested in vivo following this strategy. To date, the computer model has identified several hundred chemicals with the potential to bind AR. Empirical testing has begun, and several novel androgen antagonists have already been identified (68). The impact of these environmental chemicals on the development and health of wildlife and humans remains to be appreciated.

Acknowledgments. The authors thank Drs. L. Earl Gray, Gary Klinefelter, and Tom Wiese for their critical review of the manuscript. The authors also gratefully acknowledge Dr. L. Earl Gray for many helpful discussions regarding the development and implementation of in vivo/in vitro testing strategies for environmental endocrine-disrupting chemicals.

References

1. Kelce WR, Stone CS, Laws SC, Gray LE, Kemppainen JA, Wilson EM. Persistent DDT metabolite, *p,p'*-DDE, is a potent androgen receptor antagonist. Nature (Lond) 1995;375:581–5.

2. Kelce WR, Monosson E, Gamcsik MP, Laws SC, Gray LE Jr. Environmental hormone disruptors: evidence that vinclozolin developmental toxicity is mediated by anti-androgenic metabolites. Toxicol Appl Pharmacol 1994;126:276–85.

3. Gray LE Jr, Ostby JS, Kelce WR. Developmental effects of an environmental anti-androgen: the fungicide vinclozolin alters sex differentiation of the male rat. Toxicol Appl Pharmacol 1994;129:46–52.

4. Kelce WR. Sex steroid hormone receptor agonists and antagonists. In: Drug toxicity and embryonic development, Handbook of experimental pharmacology, Vol. 124. 1997:435–74.

5. Peterson R, Cooke P, Gray LE Jr, Kelce WR. Endocrine disruptors. In: Boekelheide K, Chapin R, eds. Reproductive and endocrine toxicology: male reproductive toxicology, Vol. 10. Sipes G, McQueen C, Gandolfi J, eds. Comprehensive toxicology. New York: Pergamon, 1997:181–91.

6. Gray LE Jr, Monosson E,and Kelce WR. Emerging issues: the effects of endocrine disruptors on reproductive development. In: Di Giulio R, Monosson E, eds. Interconnections between human and ecosystem health. New York: Chapman & Hall, 1996: 48–84.

7. Guillette LJ, Crain DA. Endocrine-disrupting contaminants and reproductive abnormalities in reptiles. Comments Toxicol 1996;5:381–99.

8. Guillette LJ, Arnold SF, McLachlan JA Ecoestrogens and embryos—is there a scientific basis for concern. Animal Reprod Sci 1996;42:13–24.

9. Toppari J, Larsen JC, Christiansen P, Giwercman A, Grandjean P, Guillette LJ, et al. Male reproductive health and environmental chemicals with estrogenic effects. Miljoprojekt 1995;290:1–166.

10. Kelce WR, Wilson EM. Clinical, functional and molecular implications of environmental antiandrogens. J Mol Med 1997;75:198–207.

11. Kavlock RJ, Daston GP, DeRosa C, Fenner-Crisp P, et al. Research needs for the risk assessment of health and environmental effects of endocrine disruptors: a report of the US EPA-sponsored workshop. Environ Health Perspect 1996;104:715–40.

12. Zhou Z-X, Lane MV, Kemppainen JA, French FS, Wilson EM. Specificity of ligand-dependent androgen receptor stabilization: receptor domain interactions influence ligand dissociation and receptor stability. Mol Endocrinol 1995;9:208–18.

13. Wong CI, Zhou ZX, Sar M, Wilson EM. Steroid requirement for androgen receptor dimerization and DNA binding: modulation by intramolecular interactions between the NH_2-terminal and steroid binding domains. J Biol Chem 1993;268:19004–12.

14. Ho KC, Marschke KB, Tan JA, Power SGA, Wilson EM, French FS. A complex response element in intron 1 of the androgen regulated 20 kDa protein gene displays cell-type dependent androgen receptor specificity. J Biol Chem 1993;268:27226–35.

15. Tan JA, Marschke KB, Ho KC, Perry ST, Wilson EM, French FS. Response elements of the androgen regulated C3 gene. J Biol Chem 1992;267:4456–66.

16. Eckert RL, Katzenellenbogen BS. Physical properties of estrogen receptor complexes in MCF-7 human breast cancer cells. J Biol Chem 1982;257:8840–6.

17. Hansen JC, Gorski J. Conformational transitions of the estrogen receptor monomer. J Biol Chem 1986;261:13990–6.

18. Martin PM, Berthois Y, Jensen EV. Binding of antiestrogens exposes an occult antigenic determinant in the human estrogen receptor. Proc Natl Acad Sci USA 1988; 85:2533–7.

19. Truss M, Bartsch J, Beato M. Antiprogestins prevent progesterone recptor binding to hormone responsive elements in vivo. Proc Natl Acad Sci USA 1994;91:11333–7.

20. Wong, CI, Kelce WR, Sar M, Wilson EM. Androgen receptor antagonist versus agonist

activities of the fungicide vinclozolin relative to hydroxyflutamide. J Biol Chem 1995;270:19998–20003.

21. Kemppainen JA, Lane MV, Sar M, Wilson EM. Androgen receptor phosphorylation, turnover, nuclear transport, and transcriptional activation. J Biol Chem 1992;267:968–74.

22. Langley E, Zhou Z-X, Wilson EM. Evidence for an anti-parallel orientation of the ligand-activated human androgen receptor dimer. J Biol Chem 1995;270:29983–90.

23. Kuil CW, Mulder E. Mechanism of antiandrogen action: conformational changes of the receptor. Mol Cell Endocrinol 1994;102:R1–5.

24. Kuil CW, Berrevoets CA, Mulder E. Ligand-induced conformational alterations of the androgen receptor analyzed by limited trypsinization. J Biol Chem 1995;270:27569–76.

25. Wilson JD. Sexual differentiation. Annu Rev Physiol 1978;40:279–306.

26. George FW, Wilson JD. Sex determination and differentiation. In: Knobil E, Neill J, eds. The physiology of reproduction. New York: Raven Press, 1988:3–26.

27. George FW. Androgen metabolism in the prostate of the finasteride-treated, adult rat: a possible explanation for the differential action of testosterone and 5α-dihydrotestosterone during development of the male urogenital tract. Endocrinology 1997;138:871–7.

28. Wilson EM, French FS. Binding properties of androgen receptors. Evidence for identical receptors in rat testis, epididymis and prostate. J Biol Chem 1976;25:5620–9.

29. Mizokami A, Chang C. Induction of translation by the 5′-untranslated region of human androgen receptor mRNA. J Biol Chem 1994;269:25655–9.

30. Lubahn DB, Joseph DR, Sullivan PM, Willard HF, French FS, Wilson EM. Cloning of human androgen receptor complementary DNA and localization to the X chromosome. Science 1988;240:327–30.

31. Brown CJ, Goss SJ, Lubahn DB, Joseph DR, Wilson EM, French FS, et al. Androgen receptor locus on the human X chromosome: regional localization to Xq 11–12 and description of a genetic polymorphism. Am J Hum Genet 1989;44:264–9.

32. Mahtani MM, Lafrenier RG, Kruse TA, Willard HF. An 18-locus linkage map of the pericentromeric region of the human X-chromosome: genetic framework for mapping X- linked disorders. Genomics 1991;10:849–57.

33. Wilson CM, Griffin JE, Wilson JD, Marcelli M, Zoppi S, McPhaul MJ. Immunoreactive androgen receptor expression in subjects with androgen resistance. J Clin Endocrinol Metab 1992;75:1474–8.

34. Quarmby VE, Kemppainen JA, Sar M, Lubahn DB, French FS, Wilson EM. Expression of recombinant androgen receptor in cultured mammalian cells. Mol Endocrinol 1990;4:1399–407.

35. Jenster G, van der Korput HAGM, van Vroonhoven C, van der Kwast TH, Trapman J, Brinkmann AO. Domains of the human androgen receptor involved in steroid binding, transcriptional activation and subcellular localization. Mol Endocrinol 1991;5:1396–404.

36. Lubahn DB, Joseph DR, Sar M, Tan J-A, Higgs HN, Larson RE, et al. The human androgen receptor: complementary deoxyribonucleic acid cloning, sequence analysis and gene expression in prostate. Mol Endocrinol 1988;2:1265–75.

37. Chang C, Kokontis J, Liao S. Structural analysis of complementary DNA and amino acid sequences of human and rat androgen receptors. Proc Natl Acad Sci USA 1988;85:7211–5.

38. Trapman J, Klaassen P, Kuiper GGJM, van der Korput JAGM, Faber PW, van Rooij

HCJ, Geurts van Kessel A, Voorhorst MM, Mulder E, Brinkmann AO. Cloning, structure, and expression of a cDNA encoding the human androgen receptor. Biochem Biophys Res Commun 1988;153:241–8.

39. Tilley WD, Marcelli M, Wilson JD, McPhaul MJ. Characterization and expression of a cDNA encoding the human androgen receptor. Proc Natl Acad Sci USA 1989;86:327–31.

40. Lubahn DB, Brown TR, Simental JA, Higgs HN, Migeon CJ, Wilson EM, et al. Sequence of the intron/exon junctions of the coding region of the human androgen receptor gene and identification of a point mutation in a family with complete androgen insensitivity. Proc Natl Acad Sci USA 1989;86:9534–8.

41. Bentvelsen FM, McPhaul MJ, Wilson JD, George FW. The androgen receptor of the urogenital tract of the fetal rat is regulated by androgen. Mol Cell Endocrinol 1994; 105:21–6.

42. Fail PA, Pearce SW, Anderson SA, Tyl RW, Gray LE. Endocrine and reproductive toxicity of vinclozolin (vin) in male Long-Evans hooded rats. Fundam Appl Toxicol 1995;15:293.

43. Szokolay A, Rosival L, Uhnak J, Madaric A. Dynamics of benzene hexachloride (BHC) isomers and other chlorinated pesticides in the food chain and in human fat. Ecotoxicol Environ Saf 1977;1:349–59.

44. Spindler M. DDT: health aspects in relation to man and risk/benefit assessment based thereupon. Residue Rev 1983;90:1–34.

45. Wyllie J, Gabica J, Benson WW. Comparative organochlorine pesticide residues in serum and biopsied lipoid tissue: a survey of 200 persons in southern Idaho—1970. Pestic Monit J 1972;6:84–8.

46. Barquet A, Morgade C, Pfaffenberger CD. Determination of organochlorine pesticides and metabolites in drinking water, human blood serum, and adipose tissue. J Toxicol Environ Health 1981;7:469–79.

47. Adamovic VM, Sokic B, Smiljanski M-J. Some observations concerning the ratio of the intake of organochlorine insecticides through the food and amounts excreted in the milk of breast-feeding mothers. Bull Environ Contam Toxicol 1978;20:280–5.

48. O'Leary JA, Davies JE, Edmundson WF, Reich GA. Transplacental passage of pesticides. Am J Obstet Gynecol 1970;107:65–8.

49. Rogan WJ, Gladen BC, McKinney JD, Carreras N, Hardy P, Thullen J, et al. Polychlorinated biphenyls (PCB's) and dichlorodiphenyl dichloroethane (DDE) in human milk: effects of maternal factors and previous lactation. Am J Public Health 1986; 76:172–7.

50. Robison AK, Schmidt WA, Stancel GM. Estrogenic activity of DDT: estrogen-receptor profiles and the responses of individual uterine cell types following o,p'-DDT administration. J Toxicol Environ Health 1985;16:493–508.

51. Imperato-McGinley J, Sanchez RS, Spencer JR, Yee B, Vaugan ED. Comparision of the effects of the 5-alpha-reductase inhibitor finasteride and the antiandrogen flutamide on prostate and genital differentiation: dose response studies. Endocrinology 1992;131:1149–56.

52. Simonich SL, Hites RA. Global distribution of persistent organochlorine compounds. Science 1995;269:1851–4.

53. Bouwman H, Cooppan RM, Becker PJ, Ngxongo S. Malaria control and levels of DDT in serum of two populations in Kwazulu. J Toxicol Environ Health 1991;33:141–55.

54. Bouwman H, Reinecke AJ, Cooppan RM, Becker PJ. Factors affecting levels of DDT

and metabolites in human breast milk from Kwazulu. J Toxicol Environ Health 1990;31:93–115.

55. Curley A, Copeland MF, Kimbrough RD. Chlorinated hydrocarbon insecticides in organs of stillborn and blood of newborn babies. Arch Environ Health 1969;19:628–32.

56. Sweet RA, Schrott HG, Kurland R, Culp OS. Study of the incidence of hypospadias in Rochester, Minn, 1940–1970, and a case-control comparision of possible etiologic factors. Mayo Clin Proc 1974;49:52–8.

57. Allera A, Herbst MA, Griffin JE, Wilson JD, Schweikert H-U, McPhaul MJ. Mutations of the androgen receptor coding sequence are infrequent in patients with isolated hypospadias. J Clin Endocrinol Metab 1995;80:2697–9.

58. Hiort O, Klauber G, Cendron M, Sinnecker GGH, Keim L, Schwinger E, et al. Molecular characterization of the androgen receptor gene in boys with hypospadias. Eur J Pediatr 1994;153:317–21.

59. Wilson JD, Griffin JE, Russell DW. Steroid 5α-reductase 2 deficiency. Endocr Rev 1993;14:577–93.

60. Bentvelsen FM, Brinkmann AO, van der Linden JETM, Schroder FH, Nijman JM. Decreased immunoreactive androgen receptor levels are not the cause of isolated hypospadias. Br J Urol 1995;76:384–8.

61. Henderson BE, Benton B, Cogsgrove M, Baptista J, Aldrich J, Townsend D, et al. Urogenital tract abnormalities in sons of women treated with diethylstilbestrol. Pediatrics 1976;58:505–7.

62. Kalloo NB, Gearhart JP, Barrack ER. Sexually dimorphic expression of estrogen receptors, but not of androgen receptors in human fetal external genitalia. J Clin Endocrinol Metab 1993;77:692–8.

63. Schardein J. Hormones and hormone antagonists. In: Chemically induced birth defects, 2nd Ed. New York: Dekker, 1993:271–339.

64. Palmer T. Regulatory requirements for reproductive toxicology: theory and practice. In: Kimmel C, Buelke-Sam J, eds. Developmental toxicology. New York: Raven Press, 1981;259–88.

65. Makris S. Proposed revisions to the testing guidelines for developmental toxicity and two-generation reproductive toxicity studies conducted under FIFRA and TSCA—a progress update. Toxicologist 1995;15:287.

66. Gray LE Jr, Ostby J, Sigmonn R, Ferrell J, Rehnberg G, Linder R, et al. The development of a protocol to assess reproductive effects of toxicants in the rat. Reprod Toxicol 1988;2:281–7.

67. Zenick H, Clegg ED, Perreault SD, Klinefelter GR, Gray LE. Assessment of male reproductive toxicity: a risk assessment approach. In: Hayes AW, ed. Principles and methods of toxicology, 3rd Ed. New York: Raven Press, 1994:937–88.

68. Waller CL, Juma BW, Gray LE, Kelce WR. Three-dimensional quantitative structure activity relationships for androgen receptor ligands. Toxicol Appl Pharmacol 1996;137:219–27.

18

Targets of Chemotherapeutic Drug Action in Testis and Epididymis

BERNARD ROBAIRE AND BARBARA F. HALES

Adverse Effects of Paternal Drug Exposure on Progeny Outcome

Paternal occupational exposures to mercury, anaesthetic gases, lead, some solvents, and pesticides are associated with an increase in spontaneous abortions (1). Paternal occupations that include painters, auto mechanics, and firemen and involve exposure to metals, solvents, and pesticides are associated with birth defects (2). Adverse effects may not be apparent at birth; increased incidences of childhood cancer are found after a number of the same exposures as for birth defects (3).

Analyses of epidemiological studies and case reports have not allowed investigators to identify specific chemical or exposure paradigms as causative for male-mediated adverse effects on progeny outcome in humans. First, the exposures are to a plethora of chemicals; with very rare exceptions, there are no data on dose, duration of exposure, or potential chemical interactions. Second, infertility and pregnancy loss are very frequent events; to demonstrate a significant increase the incidence of these events in humans requires a marked augmentation. Consequently, animal studies are essential to determine whether a drug or environmental agent administered to the male can cause adverse progeny outcome.

In contrast to the human situation, there is extensive evidence from animal studies that exposure of the male to drugs (e.g., anticancer agents) and environmental chemicals (e.g., heavy metals, solvents) will result in adverse progeny outcome. There are three major mechanisms by which exposure of males to these agents may adversely affect their progeny. The first is if the conceptus is exposed directly, by exposure of the pregnant female during mating, to a chemical present in the seminal fluid. Methadone, morphine, thalidomide, and cyclophosphamide all have adverse effects on progeny outcome by this mechanism (4–6). Previous studies from our laboratories have shown that cyclophosphamide (or its metabolites) administered to the male rat is found in seminal fluid, can be transmitted to the female partner during mating, and causes a dose-dependent increase in pre-implantation loss (6).

The second mechanism acts by modulating the hypothalamic pituitary testicular axis. Endocrine disruptors may interfere with the ability of the hypothalamic pituitary complex to communicate with the testis; p,p'-DDE, a metabolite of the pesticide DDT, has been identified as an endocrine disruptor (7). Leydig and epididymal principal cells are targeted by ethane dimethanesulphonate (EDS) (8), while Sertoli cells may be targeted by compounds such as 2,3,7,8-tetrachlorodibenzo-p-dioxin (TCDD) or hexanedione (9, 10); disruption of the functions of these cells may lead to alterations in germ cell quantity or quality. In the third mechanism, the toxicant affects the male germ cell itself, either during sperm maturation in the epididymis or during spermatogenesis in the testis.

Effects of Toxicants Are Mediated by Altering Male Germ Cells

In the Epididymis

As spermatozoa transit through the epididymis, they acquire the ability to fertilize an egg. Several proteins are added and removed from the membrane of the spermatozoa. In the nucleus, cysteine-rich protamines become progressively more tightly compacted by the formation of disulfide bonds. A few drugs have been found to affect spermatozoa during their transit through the epididymis and vas deferens. Methyl chloride induced an increase in dominant lethal mutations that was thought to be a consequence of the selective inflammatory action of this drug on the epididymis; the increase in embryo loss was reversed by an anti-inflammatory agent (11, 12). Treatment of male rats with cyclophosphamide resulted in an increase in postimplantation loss (dominant lethality) when the exposed spermatozoa originated in the head or body but not the tail of the epididymis (13).

In the Testis

Most studies on the mechanisms by which drugs or environmental chemicals administered to the male have adverse effects on progeny outcome have emphasized the male germ cell in the testis as the target of drug action. Spermatogenesis is a highly ordered and regulated process. Rat spermatogonia (stem cells) undergo five mitotic cell divisions to become spermatocytes; as spermatocytes they undergo two meiotic cell divisions to form spermatids (spermatocytogenesis). Spermatids differentiate into spermatozoa, primarily by condensing nuclear elements, shedding most of their cytoplasm, and forming an acrosome and a tail (spermiogenesis) (14). One can deduce the stage specificity of the susceptibility of the germ cells to a toxicant from the timing of exposure as the germ cells proceed through spermatogenesis (15, 16). For example, in rats an effect on progeny outcome after exposure of males to a drug or X-rays for 1 week is a consequence of an effect on spermatozoa in the epididymis (13,15).

Exposure to a toxicant 2–4 or 5–6 weeks before conception represents an effect on germ cells that were first exposed to the drug as spermatids or spermatocytes, respectively; longer exposures (7–9 weeks or more) before conception represent an effect on germ cells first exposed to the toxicant as spermatogonia (15–17). Germ cells at the spermatogonial stage are very susceptible to drugs such as procarbazine or to X-ray exposure, resulting in reduced sperm numbers and an increased percentage of morphologically abnormal surviving sperm (15, 18). Following exposure of mice to anticancer drugs such as chlorambucil, a peak in mutation yield was observed when offspring were conceived from germ cells exposed as spermatids (15). In contrast, spermatozoa were the germ cells that were maximally sensitive to the specific locus mutations induced by acrylamide monomer and the dominant lethality induced by ethylnitrosourea (15).

We and others have shown that paternal administration of cyclophosphamide, a commonly used anticancer agent, had marked dose-dependent and time-specific effects on progeny outcome (13, 16, 17, 19, 20). An increase in postimplantation loss was found after 2 weeks of chronic low-dose cyclophosphamide treatment of male rats. This postimplantation loss rose dramatically to plateau at a level dependent on drug dose by 4 weeks of treatment; this postimplantation loss was associated with exposure during spermiogenesis. Postmeiotic germ cells were also most susceptible to the induction of learning abnormalities in the progeny after paternal exposure to cyclophosphamide (21). Heritable translocations were found in mice after exposure of spermatids and spermatozoa to a single high dose of cyclophosphamide (22). A dramatic increase in preimplantation loss after 5–6 weeks of exposure of male rats to cyclophosphamide suggested that this effect was mediated primarily via drug action on spermatocytes (17, 19). Exposure of rat spermatocytes to cyclophosphamide resulted in synaptic failure, fragmentation of the synaptonemal complex, and altered centromeric DNA sequences (23). An increase in malformed and growth-retarded (weight ≤75% of the littermates) fetuses was observed after 7–9 weeks of drug treatment; the malformations observed were principally hydrocephaly, edema, and micrognathia (17). Therefore, external malformations and growth retardation were produced in progeny resulted from germ cells first exposed to cyclophosphamide as spermatogonia. Spermatogonia were reported to be at low risk for the induction of heritable translocations by cyclophosphamide (22, 24).

It is clear that the time of onset of an adverse effect of drug exposure on progeny outcome is drug specific and is dependent on the phases of spermatogenesis most susceptible to insult. The reversibility of the adverse effect of drugs on progeny outcome is also of interest. In those instances in which stem cells are affected, the insult may be irreversible. If germ cells at subsequent phases of development are most affected, the time required for reversal should mirror that for the onset of drug action. We found that the dramatic and dose-dependent increase in post-implantation loss induced by exposure of male rats to cyclophosphamide was reversed within 4 weeks after termination of drug treatment (Fig. 18.1) (25), mirroring the time course for the onset of this effect.

To determine if the adverse effects of paternal drug exposure on progeny outcome could be transmitted to a subsequent generation, males were treated with cyclophosphamide and their offspring were mated. A significant increase in mal-formations and postimplantation loss was found also in the F_2 generation (Fig.

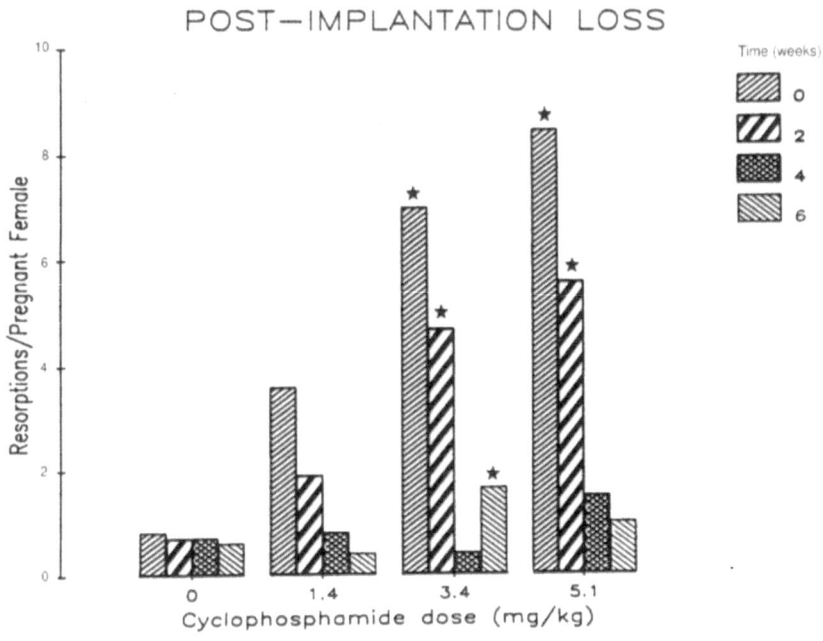

FIGURE 18.1. Reversibility of the adverse effects on progeny outcome after exposure of male rats to cyclophosphamide. Adult male rats were treated with saline (0) or cyclo-phosphamide (1.4, 3.4, or 5.1 mg kg^{-1} day^{-1}) by gavage for 9 weeks. The high level of postimplantation loss observed after 9 weeks of exposure to cyclophosphamide (O time, 3.4 or 5.1 mg kg^{-1} day^{-1} of drug) was markedly decreased by 2 weeks post treatment, and had returned to within the control range by 4 weeks post treatment. Thus, in rats, the male-mediated developmental toxicity of cyclophosphamide was reversible. Modified from Hales and Robaire (55), with permission.

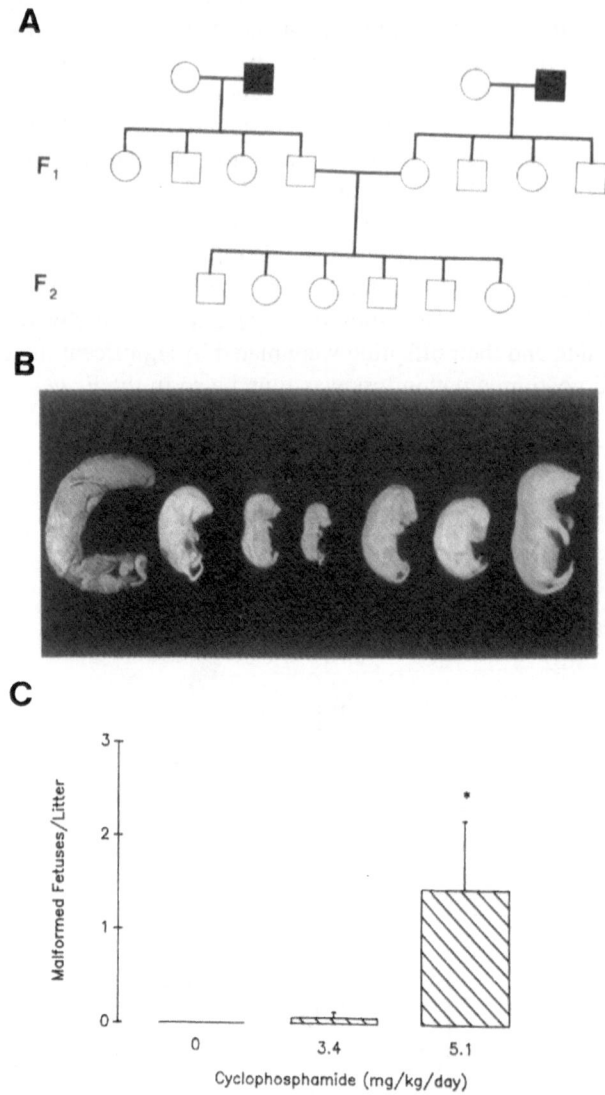

FIGURE 18.2. Heritability of the male-mediated effects of cyclophosphamide. Adult male rats were treated with saline (0) or cyclophosphamide (3.4 or 5.1 mg kg^{-1} day^{-1}) by gavage for 18 weeks. Males were mated to control females, who were allowed to deliver normally. The surviving offspring (F_1) appeared normal; they were monitored until sexual maturity, and then mated to obtain the second generation (F_2) (*A*). A significant increase in malformations was found among the offspring of rats whose sires had been exposed to cyclophosphamide (*B* and *C*). The malformations observed among the F_2 generation included gigantism, dwarfism, open eyes, omphalocele, generalized edema, and syndactyly (*B*; day 20 of gestation fetuses from the 5.1 mg kg^{-1} day^{-1} drug-treated group; the fetus on the far right is from the control group). Modified from Hales et al. (26), with permission.

18.2) (26). Malformations observed among the F_2 generation included open eyes, omphalocele, generalized edema, syndactyly, gigantism, and dwarfism. It is noteworthy that the malformations observed in the F_2 generation were similar to those found in the F_1 (17). Some of the behavioral abnormalities caused by paternal cyclophosphamide treatment also persisted in subsequent generations (27, 28).

In summary, adverse effects on progeny outcome after exposure of males to model drugs such as cyclophosphamide ranged from pre- and postimplantation effects to growth retardation, malformations, and behavioral abnormalities. Moreover, these varied adverse effects resulted from insult to germ cells at different stages; however, the spermatozoa obtained from these rats were motile and appeared grossly normal. It is important to learn how male germ cells are damaged by exposure to a drug or environmental chemical and how this damage leads to adverse reproductive outcomes.

A male germ cell with damaged DNA has two options: it may survive, repair its DNA (if possible), and go on to fertilize an oocyte, or it may undergo apoptosis. A low spontaneous incidence of apoptosis was observed in the seminiferous tubules of testes from control rats; in cyclophosphamide-exposed rats, the incidence of apoptosis increased to a level 3.5 fold above control by 12 h after treatment (29). Drug-induced apoptosis was most pronounced in premeiotic germ cells (spermatogonia and spermatocytes) in stages I–IV and XI–XIV of the seminiferous epithelium (29). Apoptosis of damaged premeiotic germ cells may serve a critical role in protecting subsequent generations from the diverse effects of toxicants.

Targets for Toxicant Action in the Male Germ Cell Nucleus

Proteins

After the spermatozoon enters the oocyte, the nucleus decondenses to restore the paternal genome to an active conformation (30). In vivo, the complete decondensation process is characterized by reduction of the disulfide bonds of the protamines, followed by their degradation and replacement with histones (31, 32). Sperm nuclei containing a lower disulfide bond content will decondense more quickly, in a hamster oocyte, and form a male pronucleus sooner (33). Spermatozoa can be decondensed in vitro by incubating them with a reducing agent (dithiothreitol) and a protease (proteinase K). We have shown that exposure to cyclophosphamide alters in vitro decondensation (34). The first phase of decondensation of spermatozoa from rats exposed to cyclophosphamide for 6 weeks was similar to control; however, in the second phase, spermatozoa from drug-

treated males remained quite compacted, whereas chromatin from control spermatozoa dispersed completely. These data could suggest that in the oocyte spermatozoa from drug-treated males may decondense and form a male pronucleus
more slowly than those from control males. However, the alkylation of protamines by cyclophosphamide may interfere with the formation of disulfide bonds
during sperm passage through the epididymis (35, 36); nuclei with a lower
disulfide bond content would decondense more quickly.

DNA

The effects of drugs on DNA in the male germ cell have been assessed using the
dominant-lethal and specific-locus mutation tests, cytogenetics, and alkaline elution. Specific-locus mutation studies have suggested that exposure of spermatogonia to chemicals or radiation yields few large lesions, while large lesions are
common after exposure of postspermatogonial germ cells. At the chromosome
level, it has proven difficult to detect even bulky deletions, aneuploidy, or
chromosomal duplications in spermatozoa using cytogenetic approaches (37).
Mature spermatozoa do not undergo mitosis. Hence, it was only by allowing
denuded hamster eggs to be fertilized by spermatozoa that chromosomal structures in the male pronucleus could be analyzed. This approach has been used to
identify the effects of age, X-irradiation, and drugs on the chromosomal banding
pattern of human sperm (38). Most recently, fluorescent in situ hybridization
(FISH) has been developed for the analysis of aneuploidy in the male genome
(39, 40).

We have assessed overall DNA damage in male germ cells after exposure of
rats to cyclophosphamide using alkaline elution (41). Under alkaline conditions,
DNA unwinds and is eluted through a filter at a rate reflecting the extent of DNA
single-stand breaks or cross-links (42). One week of treatment with cyclophosphamide caused DNA single-strand breaks that could be detected with proteinase
K in the lysis solution; no DNA cross-links were observed (41). In contrast, 6
weeks of treatment with cyclophosphamide induced a significant increase in DNA
single-strand breaks and cross-links in spermatozoal nuclei; the cross-links were
primarily the result of DNA–DNA linkages (41).

In vivo, after entering the egg and decondensing, the spermatozoon must undergo chromatin remodeling to exchange protamines with histones. It has been
suggested that this chromatin remodeling may explain why the male pronucleus is
turned on earlier than the female pronucleus (42). Transcriptional activity was
detected in the 1-cell murine embryo by G_2 when the male pronucleus was
injected with a Sp1-dependent luciferase reporter gene, but not when the female
pronucleus was injected (43). Coincidentally, there was a higher concentration of
transcription factors (Sp1 and TATA-binding protein, TBP) in the male pronucleus
than in the female (44). The DNA template function of spermatozoa from cyclophosphamide-treated male rats was determined by using an in vitro DNA synthesis system. The availability of spermatozoal DNA for template function was
not affected after 1 week of treatment with cyclophosphamide, but was markedly

affected after 6 weeks of treatment with this drug (41). At the zero time point there was a significant amount of thymidine incorporation in the chromatin from only drug-treated rat spermatozoa; this incorporation may indicate that the drug caused sufficient breaks or nicks in these nuclei to allow DNA polymerase to access the chromatin and to initiate thymidine incorporation. By 90 min, the incorporation of labeled thymidine into sperm from cyclophosphamide-treated males was nearly half of control. It is proposed that during chromatin transition processes the male genome may be in an open dynamic state with many exposed sites that are vulnerable to alkylating agents. Because there is no DNA repair during spermiogenesis, damage to the genome by alkylation at this stage may be cumulative, resulting in the production of dysfunctional germ cells.

There are "hot spots" or loci in the genome that are more susceptible to mutations. This specificity has also been observed for the visible specific-locus mutations test (45). The importance of "specific" genes as targets in mediating the adverse effects of chemicals on the male germ cells is not known. The entire male genome cannot be "shattered" because paternal genes are expressed as early as the 1-cell embryo (43), and most of the embryos survive until implantation (46). Moreover, cytogenetic studies of blastocysts sired by cyclophosphamide-treated males have shown that they have a diploid complement of chromosomes (46). Although chromosomes are not identifiable in spermatids or spermatozoa, the position of specific genes on DNA loops attached to the nuclear matrix is constant (47). Hence, it is reasonable to propose that specific gene loci may be targeted by "nonspecific" drugs, even in the absence of transcription. Thus, the paternally mediated effects of cyclophosphamide and other drugs on progeny outcome may be the consequence of gene-specific DNA damage.

The net consequence of germ cell exposure to a toxicant depends both on the extent and specificity of DNA damage and on the DNA repair processes that are active at various stages of spermatogenesis. The repair of chemically induced DNA damage in postmitotic male germ cells has been assessed with the unscheduled DNA synthesis assay, by measuring the incorporation of radiolabeled thymidine into DNA (48). Interestingly, unscheduled DNA synthesis has not been found in later-stage spermatids and spermatozoa (49), both of which are susceptible to damage leading to developmental toxicity. Few studies have assessed the presence of components of DNA repair processes in male germ cells during spermatogenesis (50). The knockout of a number of these systems (nucleotide excision repair, ERCC-1; base excision repair, DNA polymerase-β or Ref-1) in transgenic mice is lethal (51); the knockout of others (mismatch repair genes such as $Mlh1$ and $PMS2$) results in male infertility (52–54).

We propose that chronic exposure to an alkylating agent such as cyclophosphamide results in DNA damage in male germ cells at all phases of spermatogenesis. However, repair of this damaged DNA occurs principally in those germ cells that are also capable of undergoing apoptosis, that is, spermatogonia and spermatocytes. Thus, the maximal effect of such drugs is on postmeiotic germ cells, that is, elongating spermatids and spermatozoa, because they have lost the ability to repair DNA damage or to undergo apoptosis (Fig. 18.3).

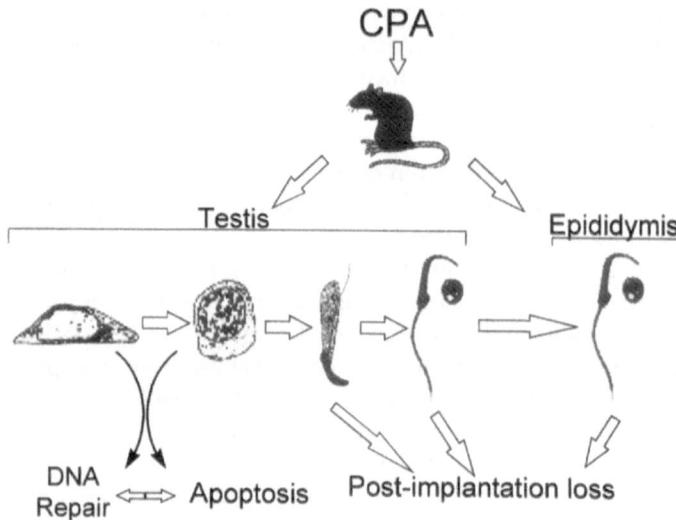

FIGURE 18.3. Schematic representation of germ cell phase specificity of the adverse effects of cyclophosphamide (CPA) treatment of male rats on progeny outcome.

Acknowledgments. Studies presented here were supported by the Medical Research Council of Canada. We express our thanks to Robert Vinson for his assistance with the preparation of Figure 18.3.

References

1. Savitz DA, Sonnenfeld NL, Olshan AF. Review of epidemiologic studies of paternal occupational exposure and spontaneous abortion. Am J Ind Med 1994;25:361–83.
2. Olshan AF, Teschke K, Baird PA. Paternal occupation and congenital anomalies in offspring. Am J Ind Med 1991;20:447–75.
3. Kuijten RR, Bunin GR, Nass CC, Meadows AT. Parental occupation and childhood astrocytoma: results of a case-control study. Cancer Res 1992;52:782–6.
4. Soyka LF, Peterson JM, Joffe JM. Lethal and sublethal effects on the progeny of male rats treated with methadone. Toxicol Appl Pharmacol 1978;45:797–807.
5. Lutwak-Mann C. Observations on the progeny of thalidomide-treated male rabbits. Br Med J 1964;1:1090–1.
6. Hales BF, Smith S, Robaire B. Cyclophosphamide in the seminal fluid of treated males: transmission to females by mating and effects on progeny outcome. Toxicol Appl Pharmacol 1986;84:423–30.
7. Kelce WR, Stone CR, Laws SC, Gray LE, Kemppainen JA, Wilson EM. Persistent DDT metabolite *p,p'*-DDE is a potent androgen receptor antagonist. Nature (Lond) 1995;375:581–5.
8. Klinefelter GR, Laskey JW, Roberts NL. In vitro/in vivo effects of ethane dimethanesulfonate on Leydig cells of adult rats. Toxicol Appl Pharmacol 1991;107:460–71.

9. Kleeman JM, Moore RW, Peterson RE. Inhibition of testicular steroidogenesis in 2,3,7,8-tetrachloro-*p*-dioxin treated rats: evidence that the key lesion occurs prior to or during pregnenolone formation. Toxicol Appl Pharmacol 1990;106:112–5.

10. Hall ES, Hall SJ, Boekelheide K. 2,5-Hexanedione exposure alters microtubule motor distribution in adult rat testis. Fundam Appl Toxicol 1995;24:173–82.

11. Chellman GJ, Bus JS, Working PK. Role of epididymal inflammation in the induction of dominant lethal mutations in Fischer 344 rat sperm by methyl chloride. Proc Natl Acad Sci USA 1986;83:8087–91.

12. Chellman GJ, Morgan KT, Bus JS, Working PG. Inhibition of methyl chloride toxicity in male F-344 rats by the anti-inflammatory agent BW755C. Toxicol Appl Pharmacol 1986;85:367–79.

13. Qiu J, Hales BF, Robaire B. Adverse effects of cyclophosphamide on progeny outcome can be mediated through post-testicular mechanisms in the rat. Biol Reprod 1992; 46:926–31.

14. Clermont Y. Kinetics of spermatogenesis in mammals: seminiferous epithelium cycle and spermatogenic renewal. Physiol Rev 1972;52:198–236.

15. Russell LB. Effects of spermatogenic cell type on quantity and quality of mutations. In: Mattison DR, Olshan AF, eds. Male-mediated developmental toxicity. New York: Plenum Press, 1994:37–48.

16. Trasler JM, Hales BF, Robaire B. Paternal cyclophosphamide treatment of rats causes fetal loss and malformations without affecting male fertility. Nature (Lond) 1985; 316:144–6.

17. Trasler JM, Hales BF, Robaire B. Chronic low dose cyclophosphamide treatment of adult male rats: effect on fertility, pregnancy outcome and progeny. Biol Reprod 1986;34:275–83.

18. Kangasniemi M, Wilson G, Parchuri N, Huhtaniemi I, Meistrich ML. Rapid protection of rat spermatogenic stem cells against procarbazine by treatment with gonadotropin-releasing hormone antagonist (Nal-Glu) and an antiandrogen (flutamide). Endocrinology 1995;136:2881–8.

19. Trasler JM, Hales BF, Robaire B. A time course study of chronic paternal cyclophosphamide treatment in rats: effects on pregnancy outcome and the male reproductive and hematologic systems. Biol Reprod 1987;37:317–326.

20. Anderson D, Bishop JB, Garner RC, Ostrosky-Wegman P, Selby PB. Cyclophosphamide: review of its mutagenicity for an assessment of potential germ cell risks. Mutat Res 1995;330:115–81.

21. Fabricant JD, Legator MS, Adams PM. Post-meiotic cell mediation of behavior in progeny of male rats treated with cyclophosphamide. Mutat Res 1983;119:185–90.

22. Generoso WM, Cattanach B, Malashenko AM. Mutagenicity of selected chemicals in mammals; the heritable translocation test. In: de Serres FJ, Shelby MD, eds. Comparative chemical mutagenesis. New York: Plenum Press, 1981:681–707.

23. Backer LC, Gibson MJ, Moses MJ, Allen JW. Synaptonemal complex damage in relation to meiotic chromosome aberration after exposure of male mice to cyclophosphamide. Mutat Res 1988;203:317–30.

24. Sotomayor RE, Cumming RB. Induction of translocations by cyclophosphamide in different germ cell stages of male mice: cytological characterization and transmission. Mutat Res 1975;27:375–88.

25. Hales BF, Robaire B. Reversibility of the effects of chronic paternal exposure to cyclophosphamide on pregnancy outcome in rats. Mutat Res 1990;229:129–34.

26. Hales BF, Crosman K, Robaire B. Increased postimplantation loss and malformations among the F_2 progeny of male rats chronically treated with cyclophosphamide. Teratology 1992;45:671–8.
27. Auroux M, Dulioust EJB, Nawar NNY, Yacoub SG, Mayaux MJ, Schwartz D, et al. Antimitotic drugs in male rat. Behavioral abnormalities in the second generation. J Androl 1988;9:153–9.
28. Auroux M, Dulioust E, Selva J, Rince P. Cyclophosphamide in the F_0 male rat: physical and behavioral changes in three successive adult generations. Mutat Res 1990;229:189–200.
29. Cai L, Hales BF, Robaire B. Induction of apoptosis in the germ cells of adult male rats after exposure to cyclophosphamide. Biol Reprod 1997;56:1490–7.
30. Naish SJ, Perreault SD, Zirkin BR. DNA synthesis in the fertilizing hamster sperm nucleus: sperm template availability and egg cytoplasmic control. Biol Reprod 1987;36:245–53.
31. Perreault SD, Naish SJ, Zirkin BR. The timing of hamster sperm nuclear decondensation and male pronucleus formation is related to sperm nuclear disulfide bond content. Biol Reprod 1987;36:239–44.
32. Perreault SD, Zirkin BR. Sperm nuclear decondensation in mammals: role of sperm-associated proteinase in vivo. J Exp Zool 1982;224:253–7.
33. Perreault SD, Barbee RR, Elstein KH, Zucker RM, Keefer CL. Interspecies differences in the stability of mammalian sperm nuclei assessed in vivo by sperm microinjection and in vitro by flow cytometry. Biol Reprod 1988;39:157–67.
34. Qiu JP, Hales BF, Robaire B. Effects of chronic low dose cyclophosphamide exposure on the nuclei of rat spermatozoa. Biol Reprod 1995;52:33–40.
35. Sega GA, Owens JG. Methylation of DNA and protamine by methyl methanesulfonate in the germ cells of male mice. Mutat Res 1983;111:227–44.
36. Sega GA, Valdivia Alcota RP, Tancongco CP, Brimer PA. Acrylamide binding to DNA and protamine of spermiogenic stages in the mouse and its relationship to genetic damage. Mutat Res 1989;216:221–30.
37. Allen JW, Collins BW, Cannon RE, McGregor PW, Afshari A, Fuscoe JC. Aneuploidy tests: cytogenetic analysis of mammalian male germ cells. In: Mattison DR, Olshan AF, eds. Male-mediated developmental toxicity. New York: Plenum Press, 1994:59–69.
38. Martin RH, Rademaker A. The effect of age on the frequency of sperm chromosomal abnormalities in normal men. Am J Hum Genet 1987;41:484–92.
39. Wyrobeck AJ, Robbins AW, Mehraein Y, Pinkel D, Weier H-U. Detection of sex chromosomal aneuploidies X-X, Y-Y, and X-Y in human sperm using two-chromosome fluorescence in situ hybridization. Am J Med Genet 1994;53:1–7.
40. Martin RH, Rademaker AW, Leonard NJ. Analysis of chromosomal abnormalities in human sperm after chemotherapy by karyotyping and fluorescence in situ hybridization (FISH). Cancer Genet Cytogenet 1995;80:29–32.
41. Qiu J, Hales BF, Robaire B. Damage to rat spermatozoal DNA after chronic cyclophosphamide exposure. Biol Reprod 1995;53:1465–73.
42. Kohn KW, Erickson LC, Ewig RAG, Friedman CA. Fractionation of DNA from mammalian cells by alkaline elution. Biochemistry 1976;15:4629–37.
43. Ram PT, Schultz RM. Reporter gene expression in G_2 of the 1-cell embryo. Dev Biol 1993;156:552–6.
44. Worrad DM, Ram PT, Schultz RM. Regulation of gene expression in the mouse oocyte

and early preimplantation embryo: developmental changes in Sp1 and TATA box-binding protein, TBP. Development (Camb) 1994;120:2347–57.

45. Favor J. Specific-locus mutation tests in germ cells of the mouse: an assessment of the screening procedures and the mutational events detected. In: Mattison DR, Olshan AF, eds. Male-mediated developmental toxicity. New York, Plenum Press: 1994:23–36.

46. Austin (Kelly) SM, Robaire B, Hales BF. Paternal cyclophosphamide exposure causes decreased cell proliferation in cleavage-stage embryos. Biol Reprod 1994;50:55–64.

47. Ward WS, Coffey DS. Specific organization of genes in relation to the sperm nuclear matrix. Biochem Biophys Res Commun 1990;173:20–5.

48. Bentley KS, Sarrif AM, Cimino MC, Auletta AE. Assessing the risk of heritable gene mutation in mammals: *Drosophila* sex-linked recessive lethal test and tests measuring DNA damage and repair in mammalian germ cells. Environ Mol Mutagen 1994;23:3–11.

49. Sotomayor RE, Sega GA, Cumming RB. Unscheduled DNA synthesis in spermatogenic cells of mice treated in vivo with the indirect alkylating agents cyclophosphamide and mitomen. Mutat Res 1978;50:229–40.

50. Walter CA, Trolian DA, McFarland MB, Street KA, Gurram GR, McCarrey JR. Xrcc-1 expression during male meiosis in the mouse. Biol Reprod 1996;55:630–5.

51. Xanthoudakis S, Smeyne RJ, Wallace FD, Curran T. The REDOX/DNA repair protein, ref-1, is essential for early embryonic development in mice. Proc Natl Acad Sci USA 1996;93:8919–23.

52. Edelmann W, Cohen PE, Kane M, Lau K, et al. Meiotic pachytene arrest in MLH1-deficient mice. Cell 1996;85:1125–34.

53. Baker SM, Plug AW, Prolla TA, Involvement of the mouse Mlh1 in DNA mismatch repair and meiotic crossing over. Nat Genet 1996;13:261–2.

54. Baker SM, Bronner CE, Zhang L, Plug AW, et al. Male mice defective in the DNA mismatch repair gene PMS2 exhibit abnormal chromosome synapsis in meiosis. Cell 1995;82:303–19.

55. Hales BF, Robaire B. The male-mediated developmental toxicity of cyclophosphamide. In: Mattison DR, Olshan AF, eds. Male-mediated developmental toxicity. New York: Plenum Press, 1994:105–16.

19

Hormonal Protection of Spermatogenic Stem Cells Against Cytotoxic Agents

MARVIN L. MEISTRICH, GENE WILSON, AND MARKO KANGASNIEMI

Various cytotoxic agents, including radiation and several chemotherapeutic drugs, produce prolonged depression of spermatogenesis in rodents and humans (1–3). However, pretreatment of rats with hormones that suppress intratesticular levels of testosterone and the completion of spermatogenesis enhances the recovery of spermatogenesis from stem cells after the cytotoxic insult (4). Although most studies of hormonal protection have employed the chemotherapeutic drug procarbazine as the cytotoxic agent, protection has also been shown following cyclophosphamide (5) and gamma irradiation (6). It had generally been assumed that the protection of spermatogenic function was a result of protecting the stem spermatogonia against killing, although alternative mechanisms such as enhancement of the number of stem cells at the time of cytotoxic treatment or of their recovery following such treatments have also been suggested (7–9).

Hormonal Protection Protocols

The timing of the hormonal procedures with respect to the administration of the cytotoxic agent depends on whether the toxicant was given as a single dose or in multiple fractions. Initial studies employed multiple (usually four) injections of procarbazine given at weekly intervals (10). The hormone treatment with testosterone was initiated several weeks before the start of the chemotherapy and continued throughout the course of chemotherapy. In attempts to better elucidate the mechanism of protection, we later decided to use only a single treatment with the cytotoxic agent (7), and in these cases the hormone treatment was given for several weeks before, and was terminated at or within one day after, the administration of the cytotoxic agent.

However, it must be noted that when depot forms of hormones are used, the termination of treatment cannot accurately be defined, and even with other methods the suppression of testosterone levels can last for some time after the cessation of hormone treatment. Although intratesticular testosterone levels were restored 1 week after cessation of administration of the gonadotropin-releasing hormone (GnRH) antagonist Nal-Glu by daily injection (11), serum testosterone levels were still partially reduced 28 days after injection of a depot form of the GnRH agonist Zoladex (12), and intratubular testosterone levels were reduced for more than 4 weeks after removal of Silastic capsules containing testosterone and estradiol (13, 14).

In a few cases, hormone treatment was given after the cytotoxic treatment. One study combined one type of hormone treatment before and during multiple injections of procarbazine and a second type of treatment after the cytotoxic treatment (15). To better elucidate the mechanism, we used single doses of the cytotoxic agent and gave hormone treatment only after cytotoxic treatment (16).

Various types of hormone treatments, all of which suppress intratesticular levels of testosterone, have been used to achieve protection of spermatogenic function. These include treatment with testosterone alone, testosterone combined with low or moderate doses of estradiol or a progestin, GnRH agonists or antagonists, or GnRH analogs combined with antiandrogens (17, 18). The employment of antiandrogens was based on the assumption that effective suppression of the spermiogenesis was a significant factor in achieving protection, but it was never determined whether the addition of the antiandrogen to the protocol resulted in greater protection of spermatogenic recovery. It should be noted that, although suppression of testosterone and spermatogenesis with the estrogen agonist/ antagonist clomiphene citrate resulted in protection of spermatogenesis from procarbazine (19), using estradiol alone did not (20).

Failure to Explain Protection in Terms of Stem Cell Survival

Because most of the studies demonstrated hormonal protection of recovery of spermatogenesis from stem cells against procarbazine (Fig. 19.1a), we originally chose that as the model toxicant to formulate hypotheses as to the mechanism of protection, which are given in Table 19.1 (8). Studies showed that there was no effect of hormone pretreatment on the systemic toxicities of procarbazine (body weight loss, decline in lymphocyte count) and even a slight sensitization of the cytotoxicity of procarbazine toward differentiating spermatogonia and early spermatocytes (8). This result rules out any possible role of changes in the systemic bioactivation, pharmacokinetics, or tissue delivery of procarbazine in the mechanism of protection of spermatogenic recovery. The protection observed against gamma irradiation (Fig. 19.1b) further indicated that such factors could not be involved (6).

The survival of cells exposed to procarbazine or gamma irradiation is dependent on several factors in common. For example, oxygen is involved in both the bioactivation of procarbazine and the production and fixation of free radical

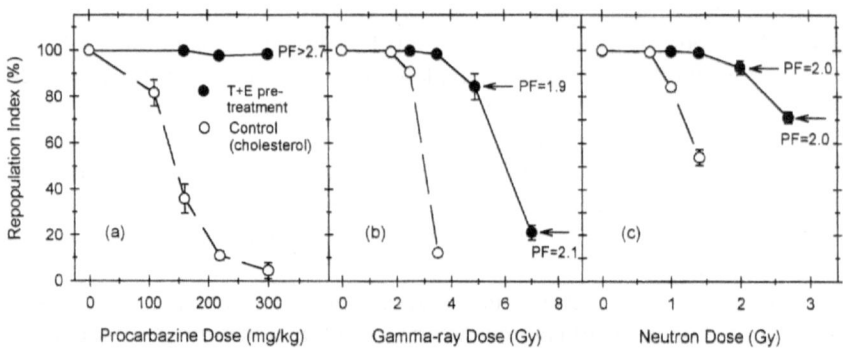

FIGURE 19.1. Effect of hormone treatment on dose–response curves for recovery of spermatogenesis, as measured by seminiferous tubule repopulation indices at 9–10 weeks after single doses of (a) procarbazine, (b) testicular gamma irradiation, or (c) testicular neutron irradiation of control or hormone-pretreated rats. Hormone treatment, 2-cm testosterone-containing and 0.5-cm estradiol-17β-containing Silastic capsules (testosterone plus estradiol, T + E) for 6 weeks before cytotoxic therapy; controls, capsules containing cholesterol. The repopulation index is the percentage of tubules containing three or more differentiating spermatogenic cells. The protection factor (PF) is the ratio of the dose required to produce a given effect in hormone-pretreated rats to the dose required to produce the same effect in controls.

TABLE 19.1. Possible mechanisms by which hormone pretreatment may modulate effects of procarbazine and other cytotoxic agents on spermatogenesis.

Mechanism of protection	Evidence against
Reduced drug metabolism and delivery to testis	Failure to protect against systemic drug toxicity or drug toxicity to differentiating germ cells or to protect against irradiation (8)
Reduced oxygen enhancement of bioactivation or radical damage, augmentation of thiol-mediated protection against radicals, or enhancement of DNA repair	Equivalent protection against neutrons as against gamma irradiation
Arrest of spermatogonial proliferation	Hormone treatment does not depress spermatogonial or spermatocyte proliferation (24)
Increased number of stem cells at time of cytotoxic treatment	Change in slope of dose–response curve for radiation (6)
	Hormone treatment does not change numbers of undifferentiated spermatogonia (24)
Enhanced recovery of spermatogenesis from surviving stem spermatogonia	Protection from genetic damage (18)

damage with gamma radiation. Thiols can protect against procarbazine by detoxification of its active metabolites and also against irradiation by reversal of radical damage in DNA. Damage produced by both procarbazine and gamma irradiation can be reversed by DNA repair processes, resulting in enhanced cell survival.

Changes in oxygen levels were unlikely to be involved in protection of stem spermatogonial recovery after procarbazine, because the adjacent differentiating spermatogonia were not protected. One argument against a role for changes in DNA repair in modulating the response of stem spermatogonia to both procarbazine and irradiation is that the types of lesions and repair processes involved are different for these two cytotoxic modalities.

To test the involvement of changes in oxygen, thiols, and DNA repair in the protection of spermatogenic recovery in hormone-treated rats, we compared the degree of protection obtained with gamma irradiation with that obtained with neutron irradiation. Neutrons produce more direct damage to DNA than the indirect free radical damage produced by gamma irradiation (21). The free radicals can be scavenged by thiols, and radical damage in DNA can be made permanent by oxygen and eliminated by thiols, whereas the direct damage produced in DNA by neutrons is unaffected by oxygen or thiols. Furthermore, although cell survival is enhanced by DNA repair processes that occur after gamma irradiation, DNA repair has little influence on cell survival after neutron irradiation. Thus, if protection of stem spermatogonia were mediated by reduction of oxygen tension, increases in thiol levels, or increased DNA repair, we would expect significantly less protection against the damage produced by neutron irradiation than by gamma irradiation.

Testes of control rats and rats treated with and testosterone plus estradiol (T+E) were irradiated with cyclotron-produced neutrons of average energy 19 MeV. The hormone pretreatment protected the testes from neutron irradiation with a protection factor of 2.0 (Fig. 19.1c), identical to that obtained with gamma rays. This result rules out any role for the interaction of oxygen, thiols, or DNA repair with the cytotoxic agents or damage produced by them in the protection of spermatogenic recovery.

Absence of Kinetic Changes in Spermatogonia

The originally proposed mechanism for hormone treatment protection of the testis was a reduction in the rate of spermatogenesis and an increase in the resistance of the resting cells (22). Arrest of spermatogonial kinetics by hormone treatment was unlikely because complete elimination of gonadotropins and nearly total ablation of androgens by hypophysectomy did not alter the kinetics of spermatocyte development (23). To directly test this mechanism, we evaluated spermatogonial proliferation in rats given identical hormone treatment to that which results in protection of the recovery of spermatogenesis after cytotoxic insult (see Fig. 19.1). Histological examination showed that the numbers and the mitotic indices of differentiating spermatogonia were unaffected by the hormone treatment (24).

Flow cytometric analysis of bromodeoxyuridine-labeled cells showed that the percentage of diploid cells undergoing DNA synthesis, the cell-cycle phase durations of the intermediate and B spermatogonia and preleptotene spermatocytes, and the progression of cells through meiotic prophase were either unchanged or very slightly increased (24). Because the numbers and kinetics of differentiating spermatogonia are unchanged by hormone treatment, their production from stem spermatogonia must also be unaffected. Although we did not directly measure the kinetics of the stem cells, they cannot be in a resting state and there is no indication that their rate of proliferation was changed.

Numbers of Stem Cells

An alternative and quite opposite mechanism was proposed in studies showing that follicle-stimulating hormone (FSH) treatment of monkeys increased the numbers of A spermatogonia and protected spermatogenesis from radiation-induced damage (25). As mentioned previously, we evaluated the numbers of spermatogonia in the rats given hormone treatment that results in protection of the recovery of spermatogenesis after cytotoxic insult (see Fig. 19.1). The numbers of A-aligned (A_{al}) spermatogonia in each of stages II–VII of the seminiferous epithelial cycle were unaffected by hormone treatment (24). As the numbers of these undifferentiated A spermatogonia were unchanged, an increase in stem cell number was unlikely. It should be noted that a large increase in number of stem cells would be required to account for the more than 20-fold increase in repopulation indices after cytotoxic treatment in hormone-treated rats (Fig. 19.1a,b). Thus, alterations of stem spermatogonial numbers cannot account for the protection of spermatogenic recovery in rats by pretreatment with gonadal steroid analogs.

Recovery of Spermatogenesis from Surviving Stem Cells

In the mouse, stem cell survival appears to be the primary determinant of the recovery of spermatogenesis, which begins immediately after cytotoxic insult and progresses until a plateau of recovery is achieved (26). In the rat and human, surviving stem cells can fail to produce sperm for prolonged periods of time; thus, the paracrine interactions necessary for the differentiation of stem cells appear to be more sensitive targets than killing of stem cells, and these are affected by lower doses (2).

After irradiation of rat testes, spermatogenesis showed a transient initial recovery but then declined with no subsequent recovery for more than 1 year. For example, after 3.5 Gy gamma irradiation, the repopulation index reached a maximum at 6 weeks after irradiation and then declined steadily (Fig. 19.2, open symbols) and was almost zero at 60 weeks (2). The nonrepopulating tubules contained normal numbers of Sertoli cells, although their appearance was somewhat abnormal. In addition, many of these tubules contained undifferentiated A

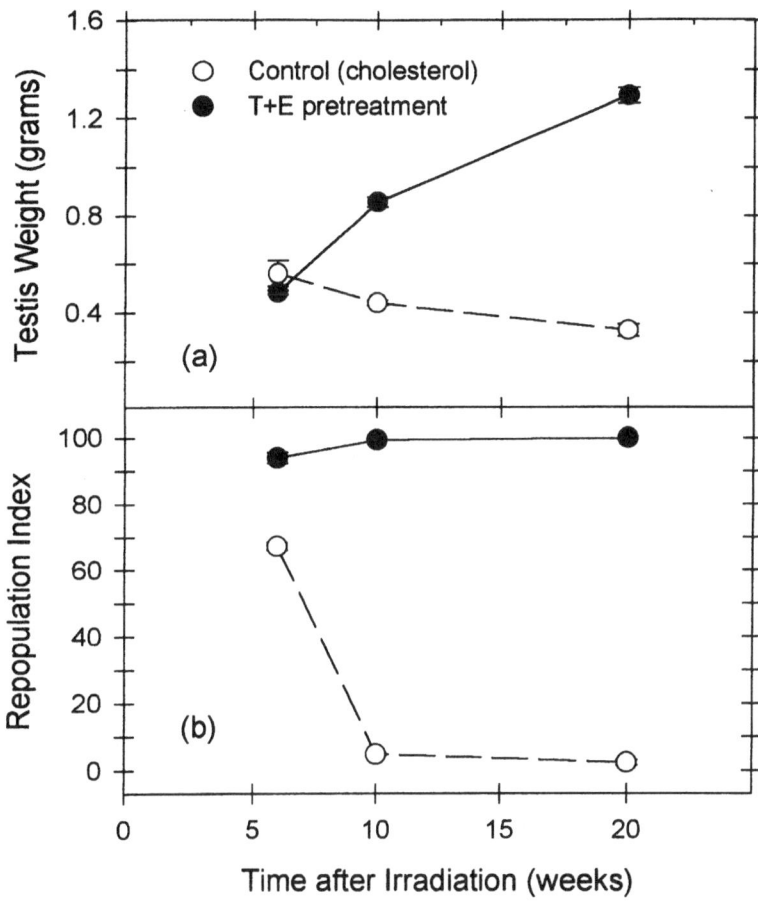

FIGURE 19.2. Effect of hormone pretreatment (T + E, 6 weeks) on the time course of recovery of spermatogenesis after 3.5 Gy of gamma irradiation as measured by (a) testicular weight and (b) repopulation index.

spermatogonia. These undifferentiated A spermatogonia were proliferating, yet their numbers remained relatively constant with time from 6 to 60 weeks after irradiation. Undifferentiated A spermatogonia have also been observed in non-repopulating seminiferous tubules after procarbazine treatment and neutron irradiation. Thus, the hormonal protection of rat spermatogenesis from the effect of cytotoxic agents can involve the regulation of spermatogenic recovery after cytotoxic treatment from these stem cells rather than the survival of the cells. That spermatogenic recovery occurred naturally after cytotoxic treatment of the mouse and was primarily limited by the survival of stem cells is consistent with the failure to obtain gonadal protection from cytotoxic damage with hormone pretreatment in the mouse (27).

The reason that surviving A spermatogonia fail to differentiate in the rat is unknown at present and likely involves the absence of appropriate paracrine growth factors in their environment. Because it is possible that such paracrine factors are regulated by endocrine alterations, we examined the levels of hormones and their receptors in the irradiated rats. Serum FSH concentrations and the amounts of testosterone per testis were slightly elevated in irradiated rats, but intratesticular testosterone concentrations were elevated fourfold at 6 weeks or more after irradiation (16). Preliminary observations indicated that FSH receptor mRNA (Tribley et al., unpublished observations) and androgen receptor (Marcelli and Meistrich, unpublished observations) were present in the irradiated testes at normal levels. Thus, the hormones and receptors responsible for Sertoli cell function are present in the irradiated testes. We considered the possibility that the high intratesticular testosterone concentrations observed after irradiation, either directly or through a metabolite of testosterone, were actually detrimental to spermatogenesis and that reduction of these levels would promote the recovery of spermatogenesis.

Hormonal Modulation of Recovery Kinetics of Spermatogenesis

Hormone Pretreatment

The previous results from our laboratory on analysis of recovery of spermatogenesis had all been done within a narrow time window of 9–12 weeks after irradiation or chemotherapy (6–9). Based on the results (see Fig. 19.2, open symbols) showing that spermatogenesis steadily declined at times after 6 weeks, we examined whether hormone pretreatment affected this decline. When the rats were treated with hormones before irradiation, not only was this decline prevented, but progressive recovery of spermatogenesis occurred (Fig. 19.2, filled symbols). Note that at 6 weeks after irradiation the repopulation index was already higher in the hormone-treated animals than in those not treated with hormones. It could be that the ability of stem cells to produce differentiating progeny in the irradiated-only testes had already declined, but that this was prevented by the hormone pretreatment.

Protection of recovery of spermatogenesis after cytotoxic damage by hormone pretreatment of the rat appears not to be a result of protection of stem cells from killing by the cytotoxic agent, but rather a result of subsequent enhancement of stimulation of stem spermatogonial differentiation by protection against mechanisms that inhibit this process. As demonstrated next, suppression of intratesticular testosterone after cytotoxic treatment similarly stimulated the recovery of spermatogenesis. Because the hormone pretreatment may continue to reduce intratesticular testosterone levels after the treatment is stopped (12–14), it may stimulate recovery of spermatogenesis by a similar mechanism. Changes in Leydig cells, including their proliferation and location within the lamina propria of tubules, have also been observed for 4 weeks following the cessation of T+E treatment (14), further indicating an altered hormonal status of the testis during this period.

Hormone Posttreatment

The recovery of spermatogenesis was also examined in rats given hormone treatment only after irradiation. Systemic treatment with testosterone, either at dose rates that highly or partially suppressed intratesticular testosterone levels, resulted in enhancement of recovery of spermatogenesis, including elimination of the decline that would normally have been observed after 6 weeks (Kurdoglu and Meistrich, unpublished results) (16). Treatment with GnRH agonists (Fig. 19.3) or antagonists (not shown) actually produced a greater recovery of spermatogenesis than did testosterone treatment. Although spermatogonial development through the spermatocyte stages is stimulated by GnRH agonist treatment, as indicated by the high repopulation index at 10 weeks (Fig. 3), the testicular weight is decreased because the GnRH agonist reduced intratesticular testosterone, inhibiting spermiogenesis as well as other aspects of testicular metabolism. However, after the cessation of GnRH-agonist treatment, complete spermatogenesis progressively recovered as indicated by increased testis weights (Fig. 19.3), numbers of late spermatids in histological sections, sperm count, and fertility (16).

It is not necessary to administer the hormone treatment immediately after irradiation. GnRH agonist treatment initiated at 20 weeks after irradiation also stimulated the repopulation of seminiferous tubules, and after its cessation, progressive recovery of spermatogenesis continued (Fig. 19.3). However, the delayed treatment was not as effective as immediate treatment. It is not known whether delayed treatment merely requires a longer time to act or if the eventual levels of recovery would still be lower than with immediate treatment. GnRH agonists can even stimulate the recovery of spermatogenesis when treatment is initiated at a time after irradiation when the testis contains no differentiating spermatogenic cells (16).

Discussion

The mechanism of the decline in the ability of type A spermatogonia to differentiate with time after 6 weeks postirradiation is not known. Because hormone treatments given either before or after cytotoxic treatments can prevent this decline and cause progressive recovery of spermatogenesis, endocrine factors must be involved. Although FSH was a candidate, the hormone treatments used here did not have a consistent or a large effect on serum FSH levels (16, 17). However, they consistently lowered intratesticular testosterone concentrations. In addition, the elevation of intratesticular testosterone concentrations after cytotoxic treatments support a role for it or one of its metabolites in the decline of spermatogenesis and as the factor modulated by the hormone treatment to stimulate the recovery. The

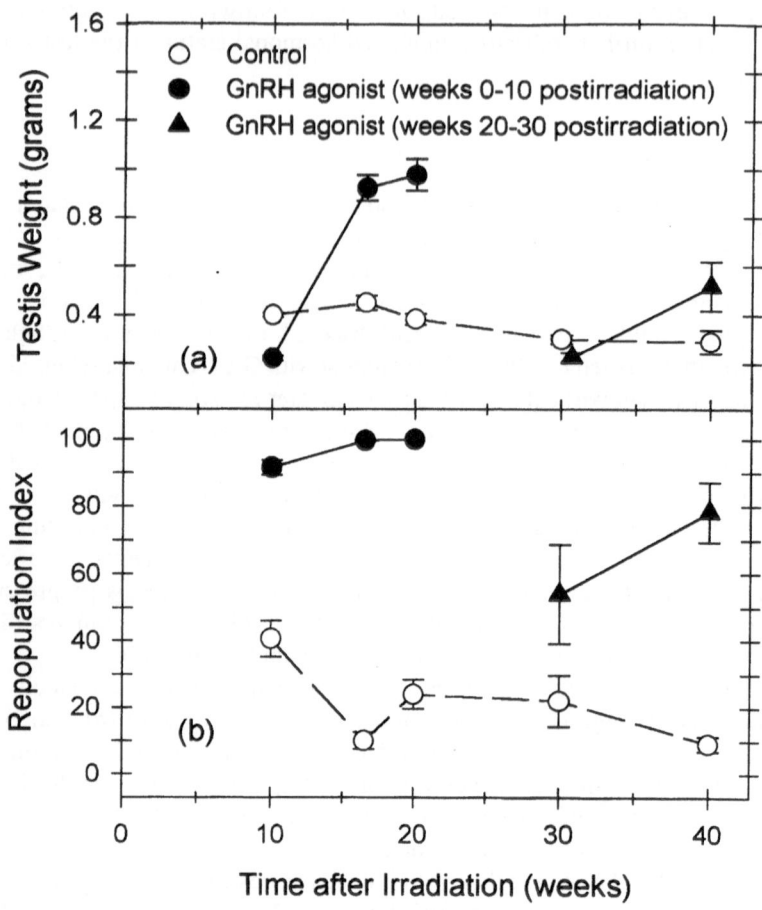

FIGURE 19.3. Effect of hormone posttreatment for 10 weeks with gonadotropin-releasing hormone (GnRH) agonist on the time course of recovery of spermatogenesis after 3.5 Gy of gamma irradiation as measured by (a) testicular weight and (b) repopulation index. GnRH agonist treatment was begun either immediately after irradiation or after a delay of 20 weeks. Controls received sham or vehicle injections. Results obtained with Zoladex (16) and Lupron were similar and are combined.

metabolite estradiol-17β should be investigated because aromatase is present in testes, Sertoli cells have estrogen receptor, and estrogens often exert effects antagonistic to those of androgens. The involvement of estradiol in inhibition of recovery of spermatogenesis would explain the failure to obtain protection with estradiol pretreatment alone (20), but not the better protection when estradiol is given in addition to testosterone (7).

Because steroid hormone receptors have not been localized in A spermatogonia, the hormonal effects are more likely mediated through paracrine interactions with other cells. Sertoli cells are the most likely source of the paracrine factors involved because they are in intimate contact with the germ cells and produce proteins, including stem cell factor, inhibin, activin, interleukin-1α, interleukin-6, seminiferous growth factor, and insulin-like growth factor-II (IGF-II), that may be capable of modulating spermatogonial proliferation and differentiation (28). One of these, stem cell factor, which interacts with c-*kit* receptors on spermatogonia, is produced in soluble and membrane-bound forms, with the latter being more important in the testis. The relative amount of mRNA for this membrane-bound form of stem cell factor was reduced in hexanedione-treated rats, which, as did irradiated rats, also showed a prolonged failure of spermatogonial differentiation despite the presence of A spermatogonia (29). Furthermore, infusion of exogenous stem cell factor for 2 weeks enhanced the proliferation of A spermatogonia in that system. It is possible that loss of the active form of stem cell factor may also be involved in the failure of spermatogenic recovery after irradiation and that hormone treatment restores this factor and thereby stimulates spermatogenic recovery. However, stem cell factor does not seem to be regulated by testosterone, although testosterone metabolites have not been studied (30).

Summary

In summary, no mechanism can be found by which hormone treatment given before administration of a cytotoxic agent can protect the survival of spermatogonial stem cells, and the effects of hormone pretreatment on the recovery of spermatogenesis after cytotoxic exposure are markedly similar to those of hormone posttreatment. We therefore conclude that hormone pretreatment "protects" rat spermatogenesis from cytotoxic damage by counteracting the as yet unknown mechanisms by which some factor, which is likely a metabolite of testosterone, alters the paracrine environment within the tubules so that spermatogonial differentiation fails. Because the human is similar to the rat in that azoospermia may last for many years after radiation or chemotherapy despite the presence of spermatogonial stem cells, hormone treatments given either before or after (or both) cancer therapy might shorten the period of azoospermia or reverse it in individuals in whom fertility may never recover.

References

1. Meistrich ML. Critical components of testicular function and sensitivity to disruption. Biol Reprod 1986;34:17–28.
2. Kangasniemi M, Huhtaniemi I, Meistrich ML. Failure of spermatogenesis to recover despite the presence of A spermatogonia in the irradiated LBNF$_1$ rat. Biol Reprod 1996;54:1200–8.
3. Meistrich ML, Sellin RV, Lipshultz LI. Gonadal dysfunction. In: DeVita VT, Hellman S, Rosenberg SA, eds. Cancer: principles and practice of oncology. Philadelphia: Lippincott, 1997:2758–73.

4. Morris ID, Shalet SM. Protection of gonadal function from cytotoxic chemotherapy and irradiation. Bailliere's Clin Endocrinol Metab 1990;4:97–118.
5. Meistrich ML, Parchuri N, Wilson G, Kurdoglu B, Kangasniemi M. Hormonal protection from cyclophosphamide-induced inactivation of rat stem spermatogonia. J Androl 1995;16:334–41.
6. Kurdoglu B, Wilson G, Parchuri N, Ye W-S, Meistrich ML. Protection from radiation-induced damage to spermatogenesis by hormone treatment. Radiat Res 1994;139:97–102.
7. Parchuri N, Wilson G, Meistrich ML. Protection by gonadal steroid hormones against procarbazine-induced damage to spermatogenic function in $LBNF_1$ hybrid rats. J Androl 1993;14:257–66.
8. Meistrich ML, Wilson G, Ye W-S, Kurdoglu B, Parchuri N, Terry NHA. Hormonal protection from procarbazine-induced testicular damage is selective for survival and recovery of stem spermatogonia. Cancer Res 1994;54:1027–34.
9. Kangasniemi M, Wilson G, Parchuri N, Huhtaniemi I, Meistrich ML. Rapid protection of rat spermatogenic stem cells against procarbazine by treatment with a gonadotropin-releasing hormone antagonist (Nal-Glu) and an antiandrogen (flutamide). Endocrinology 1995;136:2881–8.
10. Delic JI, Bush C, Peckham MJ. Protection from procarbazine-induced damage of spermatogenesis in the rat by androgen. Cancer Res 1986;46:1909–14.
11. Sinha Hikim AP, Swerdloff RS. Time course of recovery of spermatogenesis and Leydig cell function after cessation of gonadotropin-releasing hormone antagonist treatment in the adult rat. Endocrinology 1994;134:1627–34.
12. van Kroonenburgh MJPG, Beck JL, Vemer HM, Rolland R, Thomas CMG, Herman CJ. Effects of a single injection of a new depot formulation of an LH-releasing hormone agonist on spermatogenesis in adult rats. J Endocrinol 1986;111:449–54.
13. Awoniyi CA, Santulli R, Sprando RL, Ewing LL, Zirkin BR. Restoration of advanced spermatogenic cells in the experimentally regressed rat testis: quantitative relationship to testosterone concentration within the testis. Endocrinology 1989;124:1217–23.
14. Keeney DS, Sprando RL, Robaire B, Zirkin BR, Ewing LL. Reversal of long-term LH deprivation on testosterone secretion and Leydig cell volume, number and proliferation in adult rats. J Endocrinol 1990;127:47–58.
15. Pogach LM, Lee Y, Gould S, Giglio W, Huang HFS. Partial prevention of procarbazine induced germinal cell aplasia in rats by sequential GnRH antagonist and testosterone administration. Cancer Res 1988;48:4354–60.
16. Meistrich ML, Kangasniemi M. Hormone treatment after irradiation stimulates recovery of rat spermatogenesis from surviving spermatogonia. J Androl 1997;18.
17. Meistrich ML, Wilson G, Ye W-S, Thrash C, Huhtaniemi I. Relationship among hormonal treatments, suppression of spermatogenesis, and testicular protection from chemotherapy-induced damage. Endocrinology 1996;137:3823–31.
18. Velez de la Calle JF, Jegou B. Protection by steroid contraceptives against procarbazine-induced sterility and genotoxicity in male rats. Cancer Res 1990;50:1308–15.
19. Weissenberg R, Lahav M, Raanani P, Singer R, Regev A, Sagiv M, et al. Clomiphene citrate reduces procarbazine-induced sterility in a rat model. Br J Cancer 1995;71:48–51.
20. Morris ID, Bardin CW, Gunsalus G, Ward JA. Prolonged suppression of spermatogenesis by oestrogen does not preserve the seminiferous epithelium in procarbazine-treated rats. Int J Androl 1990;13:180–9.

21. Hall EJ. Radiobiology for the radiologist. 4th Ed. Philadelphia: Lippincott, 1994.
22. Glode LM, Robinson J, Gould SF. Protection from cyclophosphamide-induced testicular damage with an analogue of gonadotropin-releasing hormone. Lancet 1981;1: 1132–4.
23. Clermont Y, Harvey SC. Duration of the cycle of the seminiferous epithelium of normal hypophysectomized and hypophysectomized-hormone treated albino rats. Endocrinology 1965;76:80–9.
24. Meistrich ML, Wilson G, Zhang Y, Kurdoglu B, Terry NHA. Protection from procarbazine-induced testicular damage by hormonal pretreatment does not involve arrest of spermatogonial proliferation. Cancer Res 1997;57:1091–7.
25. van Alphen MMA, van de Kant HJ, de Rooij DG. Protection from radiation-induced damage of spermatogenesis in the rhesus monkey (*Macaca mulatta*) by follicle-stimulating hormone. Cancer Res 1989;49:533–6.
26. Meistrich ML. Effects of chemotherapy and radiotherapy on spermatogenesis. Eur Urol 1993;23:136–42.
27. Kangasniemi M, Dodge K, Pemberton AE, Huhtaniemi I, Meistrich ML. Suppression of mouse spermatogenesis by a gonadotropin-releasing hormone antagonist and anti-androgen: failure to protect against radiation-induced gonadal damage. Endocrinology 1996;137:949–55.
28. Jegou B. The Sertoli-germ cell communication network in mammals. Int Rev Cytol 1993;147:25–96.
29. Allard EK, Blanchard KT, Boekelheide K. Exogenous stem cell factor (SCF) compensates for altered endogenous SCF expression in 2,5-hexanedione-induced testicular atrophy in rats. Biol Reprod 1996;55:185–93.
30. Tajima Y, Nishina Y, Koshimizu U, Jippo T, Kitamura Y, Nishimune Y. Effects of hormones, cyclic AMP analogues and growth factors on steel factor (SF) production in mouse Sertoli cell cultures. J Reprod Fertil 1993;99:571–5.

Author Index

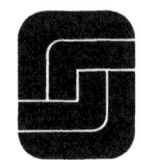

Subject Index